T0313865

The economics of vertically differentiated markets

To our beloved ones
Luca & Associates

The Economics of Vertically Differentiated Markets

Luca Lambertini

Professor of Economics, University of Bologna, Italy

and

Fellow, ENCORE, University of Amsterdam, The Netherlands

with contributions from Giorgia Bertuzzi, Chiara Celada, Luca Colombo, Giulio Ecchia, Cristina Iori, Raimondello Orsini, Gianpaolo Rossini and Carlo Scarpa

Edward Elgar

Cheltenham, UK • Northampton, MA, USA

Published by
Edward Elgar Publishing Limited
Glensanda House
Montpellier Parade
Cheltenham
Glos GL50 1UA
UK

Edward Elgar Publishing, Inc.
136 West Street
Suite 202
Northampton
Massachusetts 01060
USA

A catalogue record for this book
is available from the British Library

ISBN-13: 978 1 84542 919 5
ISBN-10: 1 84542 919 2

Printed and bound in Great Britain by MPG Books Ltd, Bodmin, Cornwall

Contents

List of Figures

List of Contributors

Giorgia Bertuzzi BA (Economics), Department of Economics, University of Bologna, Italy, giorgia.bertuzzi@libero.it

Chiara Celada BA (Economics), Department of Economics, University of Bologna, Italy, chiaracelada@libero.it

Luca Colombo PhD (Economics) Department of Economics, University of Bologna, Italy, colombo@spbo.unibo.it; and postdoctoral fellow, Trinity College Dublin, Ireland, colombol@tcd.ie

Giulio Ecchia Professor of Economics, Department of Economics, University of Bologna, Piazza Scaravilli 2, 40126 Bologna, Italy, ecchia@economia.unibo.it

Cristina Iori BA (Economics), Department of Economics, University of Bologna, Italy, i.cristina@tin.it

Luca Lambertini Professor of Economics, Department of Economics, University of Bologna, Strada Maggiore 45, 40125 Bologna, Italy, lamberti@spbo.unibo.it; and Fellow, ENCORE, University of Amsterdam, The Netherlands

Raimondello Orsini Associate Professor of Economics, Department of Economics, University of Bologna, Strada Maggiore 45, 40125 Bologna, Italy, orsini@spbo.unibo.it

Gianpaolo Rossini Professor of Economics, Department of Economics, University of Bologna, Strada Maggiore 45, 40125 Bologna, Italy, rossini@spbo.unibo.it

Carlo Scarpa Professor of Economics, Department of Economics, University of Brescia, via San Faustino 74/b, 25122 Brescia, Italy, cscarpa@eco.unibs.it

Preface

Product differentiation and its influence on consumer behaviour and the performance of firms is a core topic in the existing literature in the fields of industrial organization, international trade and economic growth alike.[1] Following Archibald, Eaton and Lipsey (1986), one usually distinguishes between address and non-address approach to product differentiation. These definitions refer, respectively, to alternative ways of modelling the substitutability between any two products being supplied in a given market. If each product's characteristics are well (and univocally) defined in the space of consumer preferences, then such characteristics constitute the product's address in the same space. This approach dates back to Hotelling (1929). If instead the degree of differentiation (or substitutability) is exogenously defined in terms of the representative consumer's preference for variety, then we are within the second approach, which stems from Chamberlin (1933), and is currently used in the version due to Dixit and Stiglitz (1977). The former approach has been widely applied in the theory of industrial organization, while the second one is a cornerstone of the new trade theory (Helpman and Krugman, 1985; Krugman, 1990) as well as the theory of endogenous growth. Additionally, a further distinction can be made within the address approach, between horizontal and vertical product differentiation. Suppose two differentiated products are supplied. If, at equal prices, demand is split (not necessarily evenly so) between the two goods, then they are said to be horizontally differentiated. Otherwise, if at equal prices all consumers prefer one to the other, then they are said to be vertically differentiated. In slightly different but substantively equivalent terms, the quality of the two products is the same in the first situation while it differs across goods in the second situation. Casual observation suggests that both

[1] This book, taking a partial equilibrium perspective, does not cover any of the themes concerning product differentiation and endogenous growth. See Grossman and Helpman (1991) and Aghion and Howitt (1998), *inter alia.*

views are empirically relevant. Indeed, more often than not, horizontal and vertical differentiation coexist, although for the sake of analytical tractability they are usually treated in isolation.[2]

This book nests into the literature on vertical differentiation, investigating several specific issues in static and dynamic models that build upon the whole body of previous contributions in order to produce some original insights. That is, the aim of the book is explicitly not one of providing the reader with a comprehensive overview of the established wisdom in the theory of vertical differentiation.[3] Rather, each chapter contains the equivalent of an original paper where, however, any new result is assessed in the context of the background literature. In order to make the exposition as homogeneous as possible, the same preference and demand structure is adopted in all chapters.[4]

The book consists of two parts. The first adopts a static approach, while the second adopts a dynamic approach. In a sense, the first part is more traditional than the second. This methodological choice has been driven by the need of modelling appropriately different issues. Whenever the static approach could be considered not to entail any significant loss of generality, it has been adopted (and conversely). The outcome, hopefully, should be useful to students taking advanced courses in industrial organization, as well as to professional researchers working in the same field.

The book sets out with a chapter where I revisit the multiproduct monopoly problem dating back to Mussa and Rosen (1978), in order to connect Spence's (1975) analysis, carried out in a single-product context, to Mussa and Rosen's results, which obtain in a setup where the product range is treated as a continuous variable and therefore the model is investigated using the tools of optimal control theory. In order to build up a connection between the two approaches, I adopt an inductive method by which I obtain the continuous model as the limiting case of the discrete one. Moreover, I

---

[2] This general rule has its relevant exceptions. See Neven and Thisse (1990) and Dos Santos Ferreira and Thisse (1996).

[3] To this regard, see Beath and Katsoulacos (1991) and Anderson, de Palma and Thisse (1992).

[4] In particular, the approach employed here to model consumer preferences dates back to Mussa and Rosen (1978) and Gabszewicz and Thisse (1979). An alternative but largely equivalent formulation is due to Shaked and Sutton (1982, 1983).

introduce into the picture the additional requirement that, in order to expand the product range, the monopolist must bear a per-product fixed cost, invariant across varieties but clearly reducing net profits as the range of varieties enlarges. This creates a tradeoff between the firm's ability to screen consumers and extract their surplus on one hand, and saving upon overall fixed costs on the other hand. This extension allows me to find a completely novel result, overlooked by the previous literature, namely that if widening the product range entails a cost of its own, then the profit-seeking monopolist may well end up supplying fewer varieties than the benevolent planner.

The second chapter, written with Raimondello Orsini, deals with the role of positional externalities in a multiproduct monopoly. Positional (or status) effects appear whenever the utility that any single individual obtains through consumption is inversely related to the number of other individuals adopting the same consumption pattern. Therefore, vertical differentiation represents a natural environment for the analysis of such externalities. To begin with, we suppose that the monopolist supplies a single variety generating a positional effect. Then, we examine the perspective where the firm introduces a second variety, characterised by a lower quality level than the first, and generating no externality. We find that it is always profitable for the monopolist to expand the product range in such a way, the reason being that the screening mechanism put in operation by the introduction of the second product more than offsets the potentially negative effect on profits caused by the availability of a lower-quality non-positional variety.

In the third chapter, written with Chiara Celada, the performance of a single-product labour-managed (LM) monopoly is evaluated against those of a profit-seeking monopolist and a benevolent social planner (as in Spence, 1975), to show that the LM monopolist will always undersupply quality as compared to both the profit-seeking monopolist and the social optimum.

In the fourth chapter, I use some well known duopoly models with vertical differentiation to describe the endogenous choice of the timing of moves in an extended game with observable delay (Hamilton and Slutsky, 1990) where firms set qualities and prices. The endogenous choice of timing refers solely to the price stage, and determines whether firms will set prices simultaneously or sequentially, while qualities are set simultaneously. First, I show that the two necessary (and sufficient, if both satisfied) conditions for sequential play to emerge at equilibrium are that both leader and follower are at least weakly better off than under simultaneous play. Second, I show that if firms can commit to their respective timing decisions, there may exist a case where the leader is not necessarily better off than in the simultaneous equilibrium. Finally, in the absence of any commitment devices, it is proved that the choice of timing is time consistent if and only if the timing stage locates between the quality stage and the price stage. If instead the timing stage precedes the quality stage, then optimal timing choices are bound to be time inconsistent.

In the fifth chapter, co-authored with Giulio Ecchia, the existence of a pure-strategy subgame perfect equilibrium in qualities and prices is investigated in a duopoly model of vertical differentiation where quality improvements require a quadratic variable cost, and firms adopt a corner solution in prices such that the poorest consumer in the market enjoys zero surplus. This setup is evaluated against the alternative cases of partial and full market coverage, where a well defined and unique equilibrium is known to exist. We show that, under the corner solution in prices, there exists a parameter range where the incentive to decrease differentiation arises for the high-quality firm, preventing firms to reach a pure-strategy duopoly equilibrium.

Chapter 6, written with Carlo Scarpa, takes a policy standpoint, in order to show that the introduction of a minimum quality standard in a vertically differentiated duopoly may have repercussions on market structure, opening the possibility of predatory behaviour. The predatory equilibrium exists independently of whether or not adjustment costs are present. Moreover, whenever predation is an equilibrium, it is selected by the risk dominance criterion.

In the seventh chapter, written with Gianpaolo Rossini, different market settings are considered in a free trade environment, where firms set quality and price or quantity while adopting different technologies. Competition in prices may give rise to asymmetric standards and leave room for the intervention of governments, via a trade policy, to make firms converge to the socially optimal equilibrium and coordinate over the same technology standard. Firms endogenously adopt a common standard either when competition is tough, owing to low consumer income, or when they compete in quantities in the market stage, thanks to a precommitted (i.e., unmodifiable) quality. In this case, the coincidence between firms' behaviour and social preferences occurs. The crucial result of the endogenous coordination over common technological standards fades away when consumers are affluent or price competition occurs.

This concludes the first part of the book. The second begins with chapter 8, written with Cristina Iori. We describe a vertically differentiated market where firms choose between activating either independent ventures leading to distinct product qualities, or a joint venture for a single quality. Then, firms interact in the market over an infinite horizon, along which they either repeat the one-shot Nash equilibrium forever, or behave collusively, depending upon the level of their individual discount factors. We prove that, in equilibrium, undertaking a joint venture goes along with collusive behaviour, while the opposite holds with independent ventures. Therefore, public policies towards R&D behaviour should be designed so as not to become inconsistent with the pro-competitive attitude characterising the current legislation on marketing practices.

Chapter 9 is co-authored with Giorgia Bertuzzi. We investigate the incentives towards product innovation and sequential quality choice in a vertically differentiated market over an infinite horizon, under the assumption of full market coverage. Two firms enter the market at different points in time, with goods characterised by different quality levels. In particular, the first entrant's quality is lower than the second entrant's, due to the fact that quality improvements take time. We consider two alternative approaches. In the first, product innovation is taken to be the outcome of a deterministic process, with the innovation taking place at an exogenously given date. In the second, innovation is stochastic and obtains at any time with a fixed probability. We show that, in both cases, the first entrant strictly prefers to supply the inferior quality since this ensures higher profits as compared to the situation where the same firm offers a high-quality good.

In chapter 10, I set out with a summary of the behaviour of a single-product monopolist, a single-product social planner and a vertically differentiated duopoly in the static approach. Then, I briefly illustrate the basic elements of optimal control and differential game theory, in order to revisit the same issues in properly dynamic terms, by assuming that quality improvements are the outcome of firms' R&D efforts over time. In particular, I analyse a differential duopoly game where firms, through capital accumulation over time, supply vertically differentiated goods. The main result is that there exists a unique steady state equilibrium which is stable in the saddle point sense, where the optimal quality ratio is 4/7, as in the static model by

Choi and Shin (1992), and, under some plausible conditions on parameters, the low-quality firm attains higher profits than the high-quality firm, contrary to the acquired wisdom based upon the static approach.

An analogous method is adopted in chapter 11, written with Luca Colombo. We investigate a differential duopoly game where each firm, through capital accumulation over time, may invest both in persuasive advertising campaigns aimed at increasing the willingness to pay of consumers and in an R&D process aimed at increasing the level of own product quality. In contrast with the conclusions reached in static models, the firm providing the market with the inferior variety may earn higher profits than the rival. In addition to this, we also show that there exists a range of parameters wherein the low quality firm commands monopoly power.

The joy of thanking all the friends and colleagues that helped my co-authors and me shape this book is tempered by the certainty that the following acknowledgments are incomplete. Of those whose names deserve an explicit mention in view of their comments and encouragement, but do not appear below, I ask forgiveness. Likewise, I assume full responsibility for any residual errors or shortcomings.

We would like to warmly thank Svend Albæk, Simon Anderson, Pierpaolo Battigalli, Emanuela Carbonara, Roberto Cellini, Vincenzo Denicolò, Egbert and Hildegard Dierker, Drew Fudenberg, Paolo Garella, Birgit Grodal, Rudolf Kerschbamer, Paul Klemperer, Massimo Marinacci, Stephen Martin, Manuela Mosca, Massimo Motta, Sougata Poddar, Dan Sasaki, Claudia Scarani, Martin Slater, Christian Schultz and Piero Tedeschi for detailed and productive comments and suggestions on single chapters. For analogous reasons, we also thank the audience at ESEM'96 (Istanbul, August 1996), XXIII EARIE Conference (University of Vienna, September 1996), Microeconomics and Game Theory Conference (University of Copenhagen, June 1997), XXIV EARIE Conference (Catholic University of Leuven, September 1997), XXV EARIE Conference (University of Copenhagen, August 1998), ASSET'98 (University of Bologna, October 1998), XIII Conference on Game Theory and Applications (University of Bologna, June 1999), and ETSG 2000 (University of Glasgow, September 2000), where several chapters were presented as independent papers.

And, last but absolutely not least, during the preparation of the final manuscript we were very fortunate in working with Emma

Gordon-Walker, Emma Meldrum, Matthew Pitman and Felicity Plester of Edward Elgar, whose constant and extremely efficient assistance allowed us to manage the project most easily.

References

1] Aghion, P. and P. Howitt (1998), *Endogenous Growth Theory*, Cambridge, MA, MIT Press.

2] Anderson, S.P., A. de Palma and J.-F. Thisse (1992), *Discrete Choice Theory of Product Differentiation*, Cambridge, MA, MIT Press.

3] Archibald, G.C., B.C. Eaton and R.G. Lipsey (1986), "Address Models of Value Theory", in J. Stiglitz and G.F. Mathewson (eds), *New Developments in the Analysis of Market Structure*, London, Macmillan.

4] Beath, J. and Y. Katsoulacos (1991), *The Economic Theory of Product Differentiation*, Cambridge, Cambridge University Press.

5] Chamberlin, E.H. (1933), *The Theory of Monopolistic Competition*, Cambridge, MA, MIT Press.

6] Choi, J.C. and H.S. Shin (1992), "A Comment on a Model of Vertical Product Differentiation", *Journal of Industrial Economics*, **40**, 229-31.

7] Dixit, A.K. and J. Stiglitz (1977), "Monopolistic Competition and Optimum Product Diversity", *American Economic Review*, **67**, 297-308.

8] Dos Santos Ferreira, R. and J.-F. Thisse (1996), "Horizontal and Vertical Differentiation: The Launhardt Model", *International Journal of Industrial Organization*, **14**, 485-506.

9] Gabszewicz, J.J. and J.-F. Thisse (1979), "Price Competition, Quality, and Income Disparity", *Journal of Economic Theory*, **20**, 340-59.

10] Grossman, G.M. and E. Helpman (1991), *Innovation and Growth in the Global Economy*, Cambridge, MA, MIT Press.

11] Hamilton, J.H. and S.M. Slutsky (1990), "Endogenous Timing in Duopoly Games: Stackelberg or Cournot Equilibria", *Games and Economic Behavior*, **2**, 29-46.

12] Helpman, E. and P. Krugman (1985), *Market Structure and Foreign Trade*, Cambridge, MA, MIT Press.

13] Hotelling, H. (1929), "Stability in Competition", *Economic Journal,* **39**, 41-57.

14] Krugman, P. (1990), *Rethinking International Trade,* Cambridge, MA, MIT Press.

15] Mussa, M., and S. Rosen (1978), "Monopoly and Product Quality", *Journal of Economic Theory,* **18**, 301-17.

16] Neven, D. and J.-F. Thisse (1990), "On Quality and Variety Competition", in J.J. Gabszewicz, J.-F. Richard and L. Wolsey (eds), *Economic Decision Making: Games, Econometrics, and Optimization. Contributions in Honour of Jacques Drèze,* Amsterdam, North-Holland.

17] Shaked, A. and J. Sutton (1982), "Relaxing Price Competition through Product Differentiation", *Review of Economic Studies,* **49**, 3-13.

18] Shaked, A. and J. Sutton (1983), "Natural Oligopolies", *Econometrica,* **51**, 1469-83.

19] Spence, A.M. (1975), "Monopoly, Quality and Regulation", *Bell Journal of Economics,* **6**, 417-29.

Part I

Statics

Chapter 1

Does monopoly undersupply product quality?

Luca Lambertini[1]

1.1 Introduction

The behaviour of a vertically differentiated monopolist has received a considerable amount of attention in the literature. The main issue at stake is whether a monopolist has any incentive to supply the same quality that would be available under perfect competition (or social planning), or to distort it so as to induce self-selection on the part of consumers. The earliest contributions (Spence, 1975; Sheshinski, 1976) deal with a single-product monopolist whose cost function is convex in quality. The main conclusions reached here (Spence, 1975) are that (i) for a given output level, quality is over or undersupplied by the monopolist as compared to social planning, depending on whether the marginal valuation of quality is above or below the average valuation of quality; if they coincide, the monopolist supplies the same quality as the social planner; and (ii) the monopolist undersupplies quality if his output is close to the socially optimal one.

Several other contributions investigate a continuous model where the monopolist supplies a range of qualities, with a technology analogous to that assumed in Spence (White, 1977; Mussa and Rosen, 1978; Itoh, 1983; Maskin and Riley, 1984; Besanko, Donnenfeld and White, 1987; Champsaur and Rochet, 1989). All of these authors emphasize that differentiation within her

[1]I thank Svend Albæk, Vincenzo Denicolò, Drew Fudenberg, Paolo Garella, Manuela Mosca, Claudia Scarani, Carlo Scarpa and Christian Schultz for useful comments and suggestions. The usual disclaimer applies.

own product range enables the monopolist to discriminate among buyers with different characteristics. In order to do so, the monopolist increases the slope of the price-quality gradient compared to the social optimum. This is achieved by offering a broader quality range than the one that would be available under social planning or perfect competition. This points to the adoption of Minimum Quality Standards to correct quality distortion (Besanko, Donnenfeld and White, 1987).

The latter statements apparently contrast with Spence's findings, according to which the difference (or coincidence) between monopoly qualities and their socially optimal levels depend upon the consumers' valuation of quality itself, and the choice of which distortion to operate, whether in the quality or the quantity dimension, is taken by the monopolist accordingly. In this chapter, I use a discrete model of multiproduct monopoly sharing the same basic features of the models employed in the above mentioned literature. I derive the continuous case as the limit of the discrete model when the number of varieties tends to infinity, and prove that

- there exists at least one case, that of a uniform consumer distribution, where the continuous model follows the rules identified by Spence;
- the continuous model lacks the crucial information as to where the marginal consumer locates at equilibrium; this produces the relevant consequence that partial market coverage is treated as a case of full market coverage of a restricted sub-population of consumers, yielding quality distortion;
- the monopolist always operates a distortion in the allocation of consumers across qualities, either by undersupplying qualities or by pricing above marginal cost (thereby restricting output) while supplying the socially optimal qualities.

Then, since Spence's conclusions hold in a continuous setting as well, the opportunity of a regulation policy based on the adoption of an MQS must be reassessed, in that the considerations put forward in the existing literature may not hold true, and there can be cases where an MQS is completely ineffective.

A departure from the commonly adopted approach to modelling the behaviour of a vertically differentiated multiproduct monopolist is then proposed, by analysing explicitly the choice of the optimal number of varieties. To my knowledge, the only existing contribution where this issue is investigated is Gabszewicz, Shaked, Sutton and Thisse (1986). In their model, however, variable costs are flat w.r.t. quality and consumer preferences are defined over quality and residual income. Their analysis nests into the literature on natural oligopolies, based upon the *finiteness property*, which cannot hold in a model with convex variable costs in the Hotelling vein (cf. Shaked and Sutton, 1983; Cremer and Thisse, 1991).

In investigating how the monopolist and the planner optimise the extension of the product range, I assume that the supply of any variety entails a fixed fee independent of the quality level characterising that variety. For any strictly positive per-product fixed cost, I reach the following conclusions:

- the monopolist produces more varieties than the planner, when both serve the whole market;
- the opposite holds when both agents price some consumers out.

These two cases arise, respectively, when the market affluence is above and below well defined thresholds. Residually, I define an intermediate range where the monopolist chooses partial market coverage while the planner serves all consumers. In such interval, there exists a level of the marginal willingness to pay at which the monopolist's and the planner's optimal ranges coincide. These conclusions produce a relevant policy implication. Whenever the number of varieties supplied under monopoly exceeds the socially optimal one, and all consumers are able to purchase a variety of the good independently of the market regime, the introduction of an MQS so as to induce the monopolist to drop some low-quality goods might be adopted by a public agency in order to enhance social welfare, provided that such a policy does not induce the monopolist to switch to partial market coverage. This measure would mitigate the distortions associated with monopoly power, by reducing both the distortion of quality and the overcrowding of the product space.

The remainder of the chapter is organised as follows. Section 1.2 contains a summary of the existing literature. The discrete model is presented in section 1.3. A discussion of the results and policy implications is in section 1.4. The role of fixed costs is investigated in section 1.5. Section 1.6 contains concluding remarks.

1.2 Preliminaries: Review of the literature

Consider a population of consumers distributed over the interval $[\underline{\theta}, \overline{\theta}]$ according to a continuously differentiable distribution function $F(\theta)$. The associated density function $f(\theta)$ is assumed to be positive everywhere over the support $[\underline{\theta}, \overline{\theta}]$. Parameter θ denotes consumer's marginal willingness to pay for quality $q \in [0, \infty)$, produced at constant unit cost $C(q)$, where $C(q)$ is twice continuously differentiable, with $C(0) = 0$; $C'(q) > 0$; and $C''(q) > 0$. Total production costs of variety q are $\Gamma = x \cdot C(q)$, where x is the output level. In the remainder of the analysis, it is assumed that the market is supplied by a single firm who is unable to observe the taste parameter θ.

Each consumer buys at most one unit of the good of quality q per period of time. A generic consumer's utility function is defined as follows:

$$U = y + V(q, \theta) , \tag{1.1}$$

where y represents consumption of all other goods. The consumer buys if net unit surplus is non-negative, i.e., if $u(\theta) = V(q, \theta) - p \geq 0$. For the moment, I assume that V is thrice continuously differentiable[2] for all q and θ, with (see Besanko, Donnenfeld and White, 1987, p. 745; Champsaur and Rochet, 1989, pp. 536-542):

Assumption 1.1 $V(0, \theta) = V(q, 0) = 0$.
Assumption 1.2 $V_q(q, \theta) > 0$; $V_{qq}(q, \theta) \leq 0$; $\partial[qV_{qq}/V_q]/\partial\theta \leq 0$.
Assumption 1.3 $V_\theta(q, \theta) > 0 \; \forall q > 0$; $V_{\theta\theta}(q, \theta) \leq 0$.
Assumption 1.4 $V_{\theta q}(q, \theta) \geq 0$; $V_{q\theta\theta}(q, \theta) \leq 0$.

A further assumption is adopted concerning the distribution function:

Assumption 1.5 $(1 - F(\theta))/f(\theta)$ *is nonincreasing in* θ.

1.2.1 The single-product monopolist

Spence (1975) and Sheshinski (1976) investigate the behaviour of a single-product monopolist facing a continuum of consumers. The firm chooses the optimal quality q of the unique variety being supplied, and the price p (or output x). The demand function for the product is

$$x = \int_{\widehat{\theta}}^{\overline{\theta}} f(\theta) d\theta , \tag{1.2}$$

where $\widehat{\theta}$ denotes the marginal willingness to pay of the marginal consumer. If $p/q > \underline{\theta}$, then $\widehat{\theta} > \underline{\theta}$, i.e., the price-quality ratio at equilibrium is such that the poorest consumer in the market is unable to buy, and *partial market coverage* obtains. Otherwise, if $p/q \leq \underline{\theta}$, then $\widehat{\theta} = \underline{\theta}$, and *full market coverage* obtains,

[2] Mussa and Rosen (1978, p.303) assume that $V(q, \theta) = \theta q$.

with $x = F(\theta)$. The monopolist's profit function is defined as $\pi = (p - C(q))x$. Consumer surplus is

$$CS = \int_{\widehat{\theta}}^{\overline{\theta}} u f(\theta) d\theta = \int_{\widehat{\theta}}^{\overline{\theta}} (V(q,\theta) - p) f(\theta) d\theta \ , \tag{1.3}$$

and social welfare is

$$SW = \pi + CS \ . \tag{1.4}$$

The following holds (see Spence, 1975, p. 419):

Proposition 1.1 *For a given output level x, the monopolist undersupplies quality compared to the social optimum if*

$$\frac{1}{x} \int_{\widehat{\theta}}^{\overline{\theta}} \frac{\partial p}{\partial q} d\theta > \frac{\partial p}{\partial q} \ ,$$

i.e., if the average valuation of quality (at the margin) is larger than the marginal valuation of quality (at the margin), and conversely.

Moreover, the tendency of a monopolist to restrict the output level compared to social planning, for a given quality, must also be accounted for. Let x^S and x^M denote the output observed under social planning (or perfect competition) and monopoly, respectively. Spence (1975, p. 421) establishes that, if x^M is near x^S, then

$$\frac{1}{x^S} \int_{\widehat{\theta}}^{\overline{\theta}} \frac{\partial p}{\partial q} d\theta > \frac{\partial p}{\partial q}(x^M) \ , \tag{1.5}$$

and consequently

$$\frac{\partial SW}{\partial q}|_{q=q^M} > 0 \ , \tag{1.6}$$

i.e., the derivative of social welfare in correspondence of the optimal monopoly quality is positive, entailing that the monopolist undersupplies quality compared to the social optimum. The reverse holds when x^M is small.[3]

Therefore, the behaviour of the monopolist is determined by the interplay between (i) his incentive to produce a suboptimal quality, according to the relationship between the marginal consumer's and the average consumer's valuation of quality, and (ii) his incentive to restrict the output level. On the one hand, the above Proposition implies that there exists a

[3]The possibility that a monopolist oversupplies product quality is further investigated in Donnenfeld and White (1988) and De Meza (1997).

class of consumers' distributions producing the coincidence between the monopolist's optimal quality and the socially efficient one, i.e., those for which $(1/x)\int_{\hat{\theta}}^{\overline{\theta}}(\partial p/\partial q)d\theta = \partial p/\partial q$ holds. This can be expected to be the case if the monopolist restricts output. On the other, the monopolist may find it profitable not to restrict output and discriminate among consumers by undersupplying quality.

The incentive to distort the quality level depends on how the surplus that the firm may appropriate varies with quality. Define

$$\overline{\pi}(q) = \max_p \pi(p,q); \quad \overline{SW}(q) = \max_p SW(p,q) = SW(p = C'(q), q) \quad (1.7)$$

and

$$\beta(q) = \frac{\overline{\pi}(q)}{\overline{SW}(q)}. \quad (1.8)$$

The slope of $\beta(q)$ determines whether quality is over or undersupplied as compared to the socially efficient level. Taking logs and differentiating w.r.t. q, one obtains

$$\frac{\beta'(q)}{\beta(q)} = \frac{\overline{\pi}'(q)}{\overline{\pi}(q)} - \frac{\overline{SW}'(q)}{\overline{SW}(q)}, \quad (1.9)$$

implying that $\beta'(q)/\beta(q) = -\overline{SW}'(q)/\overline{SW}(q)$, when $\overline{\pi}'(q) = 0$. This leads to the following Proposition (Spence, 1975, p. 421):

Proposition 1.2 *The profit-maximising monopolist undersupplies quality as compared to the social optimum if $\beta'(q) < 0$, and conversely.*

A corollary to the above result is that, when $\beta'(q) = 0$, the profit-maximising quality coincides with the socially optimal one. This is the case when the inverse demand function is linear in the output level (Spence, 1975, p. 422, fn. 7).

1.2.2 The multiproduct monopolist with a continuum of qualities

Consider now the setting where the monopolist supplies a continuum of varieties $q_i \in [0, \infty)$. Alternatively to any q_i, a consumer may purchase an outside good whose quality is normalised to zero.[4] The rationale behind this

[4] This low-end alternative is assumed, either implicitly or explicitly, by Mussa and Rosen (1978), Itoh (1983), Maskin and Riley (1984) and Besanko, Donnenfeld and White (1987). The alternative at the high-end of the product spectrum is considered only by Champsaur and Rochet (1989).

assumption lies in the fact that otherwise the derivative of the profit function w.r.t. price would be positive everywhere, seemingly entailing an infinitely high price (see Champsaur and Rochet, 1989, p. 538).[5]

Being unable to observe each consumer type, the monopolist cannot perfectly discriminate. As a result, she sets price $p(\theta)$ so as to maximize profits taking into account consumer's reaction, defined as follows:

$$q(\theta) = \arg\max_{q \geq 0} V(q,\theta) - p(\theta) \; . \tag{1.10}$$

Then, writing the price schedule as $p(\theta) = V(q(\theta),\theta) - u(\theta)$, the monopolist's problem translates into

$$\max_{q(\theta),u(\theta),\hat{\theta}} \int_{\hat{\theta}}^{\overline{\theta}} \left[V(q(\theta),\theta) - u(\theta) - C(q(\theta))\right] f(\theta) d\theta \; , \tag{1.11}$$

subject to

$$u'(\theta) = V_\theta(q(\theta),\theta) \; \forall \theta \in [\hat{\theta},\overline{\theta}] \; ; \tag{1.12}$$

$$q(\theta) \text{ is nondecreasing} \tag{1.13}$$

$u(\hat{\theta}) = 0$, and $\hat{\theta} \in [\underline{\theta},\overline{\theta}]$ is the marginal willingness to pay of the marginal consumer. This problem can be treated as an optimal control problem where $q(\theta)$ is the control variable and $u(\theta)$ is the state variable, the relevant Hamiltonian function being

$$\mathcal{H} = \left[V(q,\theta) - u - C(q)\right] f(\theta) + \mu V_\theta(q,\theta) \; , \tag{1.14}$$

where μ is the co-state variable associated with the constraint (1.12).

From the necessary first order conditions, one obtains that the optimal quality assignment is given by the following expression (see Besanko, Donnenfeld and White, 1987, p. 748; Champsaur and Rochet, 1989, p. 540):

$$\Phi(q,\theta) = V(q,\theta) - C(q) - \frac{1 - F(\theta)}{f(\theta)} V_\theta(q,\theta) \; , \tag{1.15}$$

leading these authors to state that monopoly (i) deteriorates quality for all consumers but those located at $\overline{\theta}$; and (ii) sells a larger spectrum of varieties compared to the social optimum. In particular, the profit-maximising monopolist enlarges the quality range *downwards.*[6] The generic socially optimal quality $q^S(\theta)$ is defined implicitly by the solution of the first order condition

$$V_q(q,\theta) - C'(q) = 0 \; ; \tag{1.16}$$

[5] In the next section, I am going to show that the derivative of the profit function w.r.t. price is indeed always positive under full market coverage, but this does not imply that the equilibrium price may become infinitely high.

[6] Obviously, the opposite holds if the outside good is located at the high-end of the quality spectrum.

under Assumption 1.4, $q^S(\theta)$ is indeed nondecreasing. Define the socially efficient range of product varieties as $[q_L^S(\tilde{\theta}), q_H^S(\bar{\theta})]$, where $\tilde{\theta}$ identifies the marginal consumer under social planning; and the optimal monopoly range as $[q_L^M(\hat{\theta}), q_H^M(\bar{\theta})]$. The above discussion is summarised by the following

Remark 1.1 *$q_H^S(\bar{\theta}) = q_H^M(\bar{\theta})$; $q_L^S(\tilde{\theta}) > q_L^M(\hat{\theta})$; $q^S(\theta) > q^M(\theta)$ for all $\theta \in (\hat{\theta}, \bar{\theta})$; and $\hat{\theta} \geq \tilde{\theta} \geq \underline{\theta}$.*

This is correct if the quality produced by the monopolist is nondecreasing in θ, which in turn holds if the cross partial derivative $\Phi_{q\theta}$ is positive. Otherwise, bunching consumers with different tastes onto the same variety becomes optimal (see also Lemma 1 in Besanko, Donnenfeld and White, 1987, p. 749). Finally, notice that the fourth inequality in Remark 1.1 establishes that the monopolist may restrict the output level compared to the social optimum.

1.3 The discrete model

The analysis of (i) the single-product case, and (ii) the multiproduct case, with a continuum of varieties, leads to some contradictory conclusions. On the one hand, the discrete model where a unique good is produced reveals that the quality supplied by the monopolist can be lower, equal or higher than the socially optimal quality, depending on consumers' tastes. Moreover, we should expect the monopolist to distort quality downwards as output approaches the output observed under social planning. On the other hand, the continuous model yields that the optimal quality range of a monopolist is strictly larger than the efficient product lines, and contains additional qualities located between the lower bound of the socially optimal spectrum and the outside good (Champsaur and Rochet, 1989, p. 540). Moreover, the solution of the optimal control problem does not convey a clear-cut information as to the extent of market coverage.

Here, I am going to show that Spence's (1975) results hold in the multiproduct case as well. I will prove that distortion always obtains in that the allocation of consumers across product varieties is distorted compared to the social optimum. Yet, this is not always the result of a distortion in qualities. Rather, it can be the consequence of pricing above marginal cost (or restricting output), while producing the same qualities a social planner would supply. In order to analyse the multiproduct case where quality is a discrete variable, I assume that

- $q_i \in [0, \infty)$, $i = 1, 2, 3...n$; $q_k \geq q_{k-1}$ for all k, $k - 1 \in \{1, 2, 3...n\}$;

- $V(q, \theta) = \theta q$. Consequently, net surplus is $u(\theta) = \theta q - p$, if the consumer buys; zero, if he does not buy;

- $\Gamma_i = tq_i^2 x_i$, i.e., $C_i(q_i) = cq_i^2$, with $c > 0$;

- consumers are uniformly distributed with unit density over $[\underline{\theta}, \bar{\theta}]$, with $\underline{\theta} > 0$ and $\underline{\theta} = \bar{\theta} - 1$. Hence, $f(\theta) = 1$. Observe that the uniform distribution satisfies Assumption 5.

For future reference, observe that the interval of consumers' preferred varieties is $[\underline{\theta}/(2c), \bar{\theta}/(2c)]$ (see Cremer and Thisse, 1994; Lambertini, 1997). The objective of the monopolist is to maximise w.r.t. prices and qualities

$$\Pi = \sum_{i=1}^{n} \pi_i = \sum_{i=1}^{n} (p_i - cq_i^2) x_i \tag{1.17}$$

while that of the social planner consists in maximising

$$SW = \Pi + \sum_{i=1}^{n} \int_{\theta_i}^{\theta_{i+1}} u(\theta) d\theta = \sum_{i=1}^{n} \int_{\theta_i}^{\theta_{i+1}} (\theta q_i - cq_i^2) d\theta \ , \tag{1.18}$$

where $\theta_i = (p_i - p_{i-1})/(q_i - q_{i-1})$ defines the marginal willingness to pay for quality of the consumer indifferent between varieties i and $i - 1$. All individuals for which $\theta \in (\theta_{i+1}, \theta_i)$ purchase variety i, all those for which $\theta \in (\theta_i, \theta_{i-1})$ purchase variety $i - 1$, and so on. At the upper bound of the quality range, the demand for q_n is $x_n = \bar{\theta} - (p_n - p_{n-1})/(q_n - q_{n-1})$; at the lower bound, the demand for q_1 is

$$x_1 = \frac{p_2 - p_1}{q_2 - q_1} - \text{Max}\left\{ \theta_0 = \frac{p_1}{q_1}, \underline{\theta} \right\} \ , \tag{1.19}$$

i.e., either $x_1 = (p_2 - p_1)/(q_2 - q_1) - p_1/q_1 = (p_2 - p_1)/(q_2 - q_1) - \theta_0$, under partial market coverage, or $x_1 = (p_2 - p_1)/(q_2 - q_1) - \underline{\theta}$, under full market coverage.

Several of the results that can be derived in this setting are in Lambertini (1997). The reader interested in the details of the ensuing analysis is referred to that paper.

1.3.1 Partial market coverage

When $\theta_0 = p_1/q_1 > \underline{\theta} = \overline{\theta} - 1$, the poorest consumers are excluded from consumption of any variety. The following holds:

Proposition 1.3 *For any given number n of product varieties, the profit-maximising monopolist produces the same qualities as the social planner, while supplying half the output as the social planner, both overall and for each variety. In the limit, as the number of varieties tends to infinity, the social planner serves all the market, while the monopolist serves only the upper half.*

The complete proof is in Lambertini (1997, pp. 116-118). Here, I will resume some crucial elements only. The intuition behind the result that, for any given n, equilibrium qualities are the same under both regimes lies in the fact that when the distribution is uniform and demands are linear, the average valuation of quality coincides with the marginal valuation for quality (Spence, 1975; see above),[7] and $p_i = q_i(\overline{\theta} + cq_i)/2$. On this basis, the distortion operated by the monopolist takes the usual form, i.e., an output restriction operated through the price mechanism. Given the monopoly price-output decision, it can be immediately verified that the first order condition relative to any quality q_i is the same under monopoly and social planning. Equilibrium qualities, quantities and prices are summarised as follows:

$$q_i^S = q_i^M = \frac{i\overline{\theta}}{c(2n+1)}\ , \ i = 1, 2, 3...n\ ; \tag{1.20}$$

$$X^M = \sum_{i=1}^{n} x_i^M = \frac{n\overline{\theta}}{2n+1} = \frac{X^S}{2}; \tag{1.21}$$

$$x_i^M = \frac{X^M}{n} = \frac{x_i^S}{2}\ , \ i = 1, 2, 3...n\ ; \tag{1.22}$$

$$p_i^M = \frac{i\overline{\theta}^2(2n+i+1)}{2c(2n+1)^2}\ . \tag{1.23}$$

On the basis of (1.22), the following result can be established:

Corollary 1.1 *In the monopoly regime, for any n, $x_n^M = x_n^S/2$ consumers are supplied with the same quality they would buy under social planning. All remaining consumers purchase a lower quality than under social planning. Since*

$$\lim_{n\to\infty} x_n^M = 0\ ,$$

[7] Straightforward calculations are needed to show that the same applies in the case of the triangular distribution described by $f(\theta) = 2(\overline{\theta} - \theta)$.

as the number of varieties tends to infinity the consumer indexed by $\bar{\theta}$ is the only one in a position to buy the same quality as under social planning.

As anticipated above, (1.20-1.22) hold if and only if the inequality $\theta_0 > \overline{\theta} - 1$ is satisfied. Consider the monopoly setting where a single variety is produced. In this case, $\theta_0^M(1) = 2\overline{\theta}/3$. When two varieties are produced, $\theta_0^M(2) = 3\overline{\theta}/5$; when three varieties are produced, $\theta_0^M(3) = 4\overline{\theta}/7$, and so on. In general,

Proposition 1.4 *In the monopoly regime, the marginal willingness to pay for quality of the consumer who is indifferent between the lowest quality and the outside good is $\theta_0^M(n) = (n+1)\overline{\theta}/(2n+1)$. Under social planning, it is $\theta_0^S(n) = \theta_0^M(n)/2$.*

Proof. In order to prove the first statement in the above Proposition, two alternative routes can be taken. The first consists in a simple argument by induction. Note first that $\theta_0^M(n) = \theta_0^M(1)$ if a single good is supplied. Then, observe that, in the case of $n+1$ varieties, one obtains $\theta_0^M(n+1) = ((n+1)+1)\overline{\theta}/(2(n+1)+1)$; defining $\nu = n+1$, the former expression becomes $\theta_0^M(\nu) = (\nu+1)\overline{\theta}/(2\nu+1)$, which differs from $\theta_0^M(n)$ only for the presence of ν in place of n. The second consists in deriving the location of the marginal consumer from (1.20-1.23) when $i = 1$. Then, the proof that $\theta_0^S(n) = \theta_0^M(n)/2$ follows from the straightforward comparison between total outputs in (1.21). ■

Therefore, I can also state

Corollary 1.2 *Under monopoly (respectively, social planning), a necessary and sufficient condition for partial market coverage to obtain is $\theta_0^M(n) > \underline{\theta}$, (respectively, $\theta_0^S(n) > \underline{\theta}$) i.e., $\overline{\theta} < (2n+1)/n$ (respectively, $\overline{\theta} < (2n+1)/(2n)$).*

Consider the monopoly regime. The sufficiency relates, obviously, to the condition that the profits associated with partial market coverage be larger than the profits associated with full market coverage. A simple argument suffices to prove this. Consider that profit maximisation requires the choice by the monopolist of the price-quality ratio $\theta_0^M(n) = p_1/q_1$ defining the location of the marginal consumer over the interval $[\underline{\theta}, \overline{\theta}]$. If profit maximisation w.r.t. prices and quantities yields $\theta_0^M(n) > \underline{\theta}$, this implies that the price-quality schedule chosen by the monopolist is indeed optimal if and only if

$\overline{\theta} < (2n+1)/n$. If the latter inequality is not satisfied, the monopolist must take into account that the market is so rich that no consumer can be priced out, i.e., full market coverage is to be expected from the outset. As to the behaviour of the social planner, notice that the critical threshold of $\overline{\theta}$ below which the planner prices out some consumers is half the monopolist's critical threshold. The policy implications of this result are discussed in section 1.4.

1.3.2 Full market coverage

Suppose all consumers are in a position to buy, so that $X^S = X^M = F(\theta) = 1$. Demands are defined as in the previous subsections (see 1.17-1.18), with $x_1 = (p_2 - p_1)/(q_2 - q_1) - \underline{\theta}$. Moreover, $p_1 = \underline{\theta} q_1$. The following holds:

Proposition 1.5 *As long as the number of varieties is finite, the monopolist undersupplies all qualities compared to the social optimum. As the number of varieties tends to infinity, the highest quality coincides with the socially optimal one, while the difference between the lowest monopoly quality and the socially optimal one is increasing in the number of varieties and, in the limit, is equal to the range of consumers' preferred qualities.*

Again, the complete proof is in Lambertini (1997, pp. 112-116). The lower and upper bounds of the profit-maximising quality spectrum are

$$q_1^M = \frac{n(\overline{\theta} - 2) + 1}{2cn} \; ; \; q_n^M = \frac{n\overline{\theta} - 1}{2cn} \, , \tag{1.24}$$

while the socially optimal bounds are

$$q_1^S = \frac{2n(\overline{\theta} - 1) + 1}{4cn} \; ; \; q_n^S = \frac{2n\overline{\theta} - 1}{4cn} . \tag{1.25}$$

As a result, the degrees of differentiation in the two regimes are

$$q_n^M - q_1^M = \Delta q^M = \frac{n-1}{cn} \; ; \; q_n^S - q_1^S = \Delta q^S = \frac{n-1}{2cn} \, , \tag{1.26}$$

with $\Delta q^M = 2\Delta q^S$, and

$$\lim_{n\to\infty} \Delta q^M = \frac{1}{c} ; \; \lim_{n\to\infty} \Delta q^S = \frac{1}{2c} . \tag{1.27}$$

Hence, for a given output level, the distortion observed in the equilibrium quality levels under monopoly increases as one moves downwards along the quality spectrum. It is easily shown that

$$x_i^M = x_i^S = \frac{1}{n} . \tag{1.28}$$

As a result, given that the output level of any variety is the same in both regimes, undersupplying quality allows the monopolist to extract more surplus from rich consumers. In the limit, the result of "no distortion at the top" obtains:

$$\lim_{n\to\infty}(q_n^S - q_n^M) = 0; \ \lim_{n\to\infty}(q_1^S - q_1^M) = \frac{1}{2c}, \tag{1.29}$$

while all qualities lower than q_n^M become more distorted as n increases.

Notice that the solution of the monopoly problem when full market coverage is assumed from the outset implies solving $n-1$ first order conditions w.r.t. prices, since the price of the lowest quality is $p_1 = \underline{\theta} q_1$. Although known from the outset, this piece of information must be used only after writing the first order conditions concerning the $n-1$ products above q_1. Moreover, $p_1 = \underline{\theta} q_1$ coincides with the price charged on the lowest quality under partial market coverage if and only if the location of the marginal consumer under partial coverage coincides with the lower bound of the support of consumers' distribution, i.e., $\theta_0 = \underline{\theta} = \overline{\theta} - 1$. Finally, observe that, in the papers analysing the continuous model, the information that the surplus enjoyed by the marginal consumer in equilibrium must be nil is used from the outset, which *ex post* is indeed correct independently from the extent of market coverage, but is appropriately used *ex ante* only if full market coverage is expected to arise at equilibrium. Its use *ex ante* under partial market coverage eliminates one degree of freedom and modifies first order conditions, in that it transforms the optimisation problem under partial coverage into an optimisation problem under full coverage of a subset $[\hat{\theta}, \overline{\theta}]$ of the population of consumers. This procedure yields that the monopolist *always* undersupplies all qualities, and prevents from explicitly locating the marginal consumer along the support, thereby inducing the result that the monopolist may or may not restrict the extent of market coverage as compared to social planning or perfect competition.

1.4 Discussion and policy implications

I am now in a position to discuss the implications of the results derived in the previous section. The comparison between monopoly and social planning yields the following:

Proposition 1.6 *For all $\overline{\theta} \geq (2n+1)/n$, full market coverage is observed irrespective of the market regime. For all $\overline{\theta} < (2n+1)/(2n)$, partial market coverage is observed irrespective of the market regime. For* *all* $\overline{\theta} \in [(2n+1)/(2n), (2n+1)/n)$, *the market is fully covered by the social planner while it is only partially covered by the monopolist.*

When partial market coverage emerges at the monopoly equilibrium, qualities are the same as under social planning, i.e., they are undistorted. However, a distortion in the allocation of consumer across qualities is observed, due to the fact that the price-quality gradient increases in the quality level and the output level is reduced as compared to social planning. Conversely, under full market coverage, the output level is not restricted, both overall and for any single variety, and the allocation of consumers across qualities is distorted by undersupplying each quality, except the top one in the limit, when the number of varieties tends to infinity and consequently the product spectrum becomes continuous. Therefore, as the quality range becomes continuous, the result of 'no distortion at the top' is common to both settings. As long as quality is discrete, i.e., for any finite value of n, there exists a group of consumers (those identified by $\theta \in (\theta_n, \overline{\theta}]$) that, under partial market coverage, are able to buy the same quality irrespective of the firm's objective function, although they obviously pay different prices in the two cases. This can never happen under full market coverage, if quality is a discrete variable. To sum up, I can state

Proposition 1.7 *Consider any finite* n. *Then,*

(i) if $\overline{\theta} < (2n+1)/n$, *partial market coverage obtains and the monopolist supplies the socially optimal qualities and distorts the allocation of consumers across qualities through the price vector, for all values of* θ *except those in the interval* $(\theta_n, \overline{\theta}]$;

(ii) if $\overline{\theta} \geq (2n+1)/n$, *full market coverage obtains and the monopolist undersupplies all qualities, while producing the same output as the social planner for any variety.*

Proposition 1.8 *As* n *tends to infinity and the quality range becomes continuous, partial market coverage obtains if* $\overline{\theta} < 2$. *Otherwise, the market is fully covered. In both cases, there exists a unique consumer, located at* $\overline{\theta}$, *purchasing the same quality as under social planning.*

A few additional remarks are in order. First, if the market is relatively poor, the monopolist finds it optimal to restrict the output level, while the quality spectrum coincides with the socially efficient one. This implies that the misallocation of consumers is operated by the usual tendency on the part of the monopolist to price above marginal cost. The distortion in qualities emerges only if the market is so affluent that no consumer can be profitably priced out. Second, since we can imagine that, in real-world situations, the presence of fixed costs in production prevents the quality range from becoming continuous, then we may reasonably expect to observe the first case rather than the second, when the marginal valuation of quality coincides with the average valuation of quality. This issue is tackled in the next section. Third, the above analysis has relevant implications concerning the possibility of regulating the monopolist's behaviour through the adoption of a Minimum Quality Standard (MQS). This policy is investigated in the continuous setting by Besanko, Donnenfeld and White (1987). In the light of the above discussion, the MQS is ineffective under partial market coverage, when qualities are not distorted. It can be used to raise the average quality available in the market under either full or partial market coverage, provided the marginal valuation of quality is lower than the average. Under the assumption that the lower bound of the monopolist's product range is lower than the MQS, Besanko, Donnenfeld and White (1987, Proposition 1, p. 750), find that the consumers for whom the MQS is not binding purchase the same quality as in the unregulated equilibrium, while those for whom the MQS is indeed binding receive a higher quality. Moreover, some consumers may be excluded from the market after the adoption of an MQS. The first part of the statement is not generally true, in that the MQS induces a change in the prices of all varieties and a consequent modification in the assignment of consumers across qualities;[8] the second part of the statement can be true, but it implies that the MQS policy might be discontinued if partial market coverage obtains.

Finally, a relevant policy implication of Proposition 1.6 is the following:

Corollary 1.3 *If* $\overline{\theta} \in [(2n+1)/(2n), (2n+1)/n)$, *an MQS can be adopted only in combination with a price regulation such that the monopolist covers the entire market, i.e.,* $p_1 = \theta_0 q_1$.

This highlights that, used in isolation, an MQS policy may not be viable, in that whenever the distortion operated by the monopolist takes the usual form of a price increase (or output restriction) rather than a reduction in quality, authorities should rather adopt policy measures explicitly tailored on output or price, such as price caps or, perhaps, rate of return regulations. Further considerations on the feasibility of MQS regulation will be proposed at the end of the next section.

[8] In Besanko, Donnenfeld and White (1988), the population of consumers consists in two disjoint groups. In that case, it is indeed true that the introduction of the MQS affects only the quality supplied to the low-income group of consumers.

1.5 Fixed costs and the optimal product range

So far, I have assumed that the production of any given variety entails no costs other than the convex unit cost $C_i(q_i) = cq_i^2$. In this section, I will extend the analysis to account for the presence of a per-product fixed cost $\xi > 0$, constant across varieties.[9] As in the foregoing analysis, I will partly rely on results derived in Lambertini (1997).

1.5.1 Partial market coverage

Under partial market coverage, the monopolist's per-product equilibrium profits are:

$$\pi_i^M = \frac{i\overline{\theta}^3(2n-i+1)}{2c(2n+1)^3} - \xi \; \forall \, i \in \{1, 2, ...n\} \; . \tag{1.30}$$

Summing up per-product profits (1.30) over index i, I obtain the following total monopoly profits:

$$\Pi^M = \sum_{i=1}^{n} \pi_i^M = \frac{n\overline{\theta}^3(n+1)}{6c(2n+1)^2} - n\xi \; . \tag{1.31}$$

Then, differentiating (1.31) w.r.t. n, I can write the first order condition for profit maximization w.r.t. n :[10]

$$\frac{\partial \Pi^M}{\partial n} = \frac{\overline{\theta}^3 - 6c\xi[1+6n+12n^2+8n^3]}{6c(2n+1)^3} = 0 \; , \tag{1.32}$$

yielding $n_{pmc}^M = \overline{\theta}/[2(6c\xi)^{1/3}] - 1/2$ as the optimal number of varieties in the monopoly equilibrium. Subscript *pmc* stands for *partial market coverage*. In order to compare n_{pmc}^M with the socially optimal number of products, observe that, when $\xi = 0$, we have $SW^M = 3\Pi^M/2$ and $SW^S = 4SW^M/3 = 2\Pi^M$ for all n. Consequently, when n varieties are supplied at the fixed cost of ξ each, net social welfare is

$$SW^S = \frac{n\overline{\theta}^3(n+1)}{3c(2n+1)^2} - n\xi \; . \tag{1.33}$$

[9] Parameter ξ can be considered as the cost of product innovation associated with an R&D technology with constant returns. In line of principle, one may also imagine that the monopolist activates a portfolio of innovation, including cost-reducing ones. This view is investigated in Lambertini and Orsini (2000).

[10] Second order conditions are met here as well as in the remainder of this section. They are omitted for the sake of brevity.

Differentiating (1.33) w.r.t. n and solving, yields $n_{pmc}^{S} = \overline{\theta}/[2(3c\xi)^{1/3}] - 1/2$. It is immediately verified that $n_{pmc}^{M} < n_{pmc}^{S}$ over the whole parameter space $\{\overline{\theta}, \xi, c\}$, as long as ξ is strictly positive. In the presence of fixed costs, the private incentive towards product proliferation is weaker than the social incentive. This contradicts the established wisdom based on the continuous model without fixed costs, from which it emerges that "the monopolist may sell some qualities which do not even appear under competition" (Mussa and Rosen, 1978, p. 315). Obviously, as the fixed cost goes to zero, the optimal number of varieties becomes infinitely large under both monopoly and social planning.

It is worth observing that the number of varieties offered in equilibrium under both regimes increases in the affluence of the market, measured by $\overline{\theta}$. In Gabszewicz, Shaked, Sutton and Thisse (1986), the optimal number of products increases as consumers' incomes become more dispersed. Moreover, consider the conditions under which at least one variety is supplied. Under monopoly, $n_{pmc}^{M} = 1$ if $\overline{\theta} = 3(6\xi c)^{1/3}$; under social planning, $n_{pmc}^{S} = 1$ if $\overline{\theta} = 3(3\xi c)^{1/3}$. Hence, we have the following:

Proposition 1.9 *Under partial market coverage,*

(i) if $\overline{\theta} \in [1, 3(3\xi c)^{1/3})$, then neither the planner nor the monopolist find it convenient to serve the market at all;

(ii) if $\overline{\theta} \in [3(3\xi c)^{1/3}, 3(6\xi c)^{1/3})$, then the market is supplied under social planning but not under monopoly.

In correspondence of the above critical values of $\overline{\theta}$, the producer (i.e., the monopolist or the social planner, alternatively), bunch all consumers onto the only quality being supplied.

1.5.2 Full market coverage

The case of full market coverage can now be quickly dealt with. Per-product profits are

$$\pi_i^M = \frac{n\overline{\theta}((\overline{\theta} - 2) + 2n(2i - 1) - 2i(i - 1) - 1}{4n^3 c} - \xi \;\forall\, i \in \{1, 2, ...n\} \,. \quad (1.34)$$

The corresponding total monopoly profits amount to:

$$\Pi^M = \sum_{i=1}^{n} \pi_i^M = \frac{n^2\left(3\overline{\theta}^2 - 6\overline{\theta} + 4\right) - 1}{12n^2 c} - n\xi \,. \quad (1.35)$$

Social welfare under planning is

$$SW^S = \frac{4n^2\left(3\overline{\theta}^2 - 3\overline{\theta} + 1\right) - 1}{48n^2c} - n\xi \,. \tag{1.36}$$

Straightforward calculations yield the optimal number of varieties under monopoly, $n^M_{fmc} = 1/(6\xi c)^{1/3}$, and social planning, $n^S_{fmc} = 1/[2(3\xi c)^{1/3}]$, with $n^M_{fmc} > n^S_{fmc}$ over the whole parameter space $\{\overline{\theta}, \xi, c\}$, as long as ξ is strictly positive. Subscript fmc stands for *full market coverage*. As in the partial coverage case, $\lim_{\xi\to 0} n^M_{fmc} = \lim_{\xi\to 0} n^S_{fmc} = \infty$.

The above result can be given the following interpretation. When full market coverage obtains under both regimes, the monopolist finds it optimal to distort quality downwards *and* to stretch the product range below the lower bound of the planner's optimal range, in order to extract more surplus from customers, in particular the richer ones. This reinforces the well known result, dating back to Dupuit (1854; see also Ekelund, 1970), according to which the distortion affecting monopoly quality is aimed at inducing self-selection among consumers.

Observe that, under full market coverage, the number of varieties does not depend upon consumers' marginal willingness to pay for quality. The collapse of the product range into a unique quality obtains at $\xi = 1/(6c)$ (with $n^M_{fmc} = 1$) and $\xi = 1/(24c)$ (with $n^S_{fmc} = 1$), irrespective of market affluence. When $\xi > 1/(6c)$ (respectively, $\xi > 1/(24c)$), the monopolist (resp., planner) does not produce any variety.

1.5.3 Partial vs full market coverage

From the previous sections (cf., in particular, Proposition 1.6), we know that, given a generic number of varieties, (i) if $\overline{\theta} \geq (2n+1)/n$, full market coverage obtains under both market regimes; (ii) if $\overline{\theta} < (2n+1)/(2n)$, partial market coverage obtains, irrespective of the market regime; (iii) if $\overline{\theta} \in [(2n+1)/(2n), (2n+1)/n)$, the planner chooses full market coverage while the monopolist chooses partial market coverage. When optimal product ranges are endogenously determined in the presence of a positive and finite per-product fixed cost ξ, we observe what follows.

First, when $n^S_{pmc} = n^S_{fmc}$, the planner switches from partial to full market coverage. This happens at $\overline{\theta} = (2n^S_{pmc} + 1)/(2n^S_{pmc}) = 1 + (3\xi c)^{1/3}$. Second, when $n^M_{pmc} = n^M_{fmc}$, the monopolist switches from partial to full market coverage. This happens at $\overline{\theta} = (2n^M_{pmc} + 1)/(n^M_{pmc}) = 2 + (6\xi c)^{1/3}$. Clearly, the latter level of $\overline{\theta}$ is larger than the former. Hence, the optimal policy of the planner and the monopolist, respectively, can be characterised as follows:

- $\overline{\theta} \in \left[1, 1 + (3\xi c)^{1/3}\right)$. Over this range, both agents choose partial market coverage, with $n^M_{pmc} < n^S_{pmc}$.

- $\overline{\theta} > 2 + (6\xi c)^{1/3}$. In this region, both agents fully cover the market, with $n^M_{fmc} > n^S_{fmc}$.

- In the intermediate region where $\overline{\theta} \in \left(1 + (3\xi c)^{1/3}, 2 + (6\xi c)^{1/3}\right)$, the monopolist chooses partial market coverage while the planner chooses full market coverage, and we have to evaluate n^M_{pmc} against n^S_{fmc}.

To this regard, it is quickly established that

Proposition 1.10 *If* $\overline{\theta} \in \left(1 + (3\xi c)^{1/3}, 6^{1/3}(3^{2/3} + 3(\xi c)^{1/3})/3\right)$, *then* $n^M_{pmc} < n^S_{fmc}$. *Conversely, if* $\overline{\theta} \in \left(6^{1/3}(3^{2/3} + 3(\xi c)^{1/3})/3, 2 + (6\xi c)^{1/3}\right)$, *then* $n^M_{pmc} > n^S_{fmc}$.

The discussion I have just carried out can be summarised in the following terms. The monopolist's product range is wider than the planner's when both agents fully cover the market. This stems from the combination of the distortion observed in the quality level and the overcrowding observed in the product space. The opposite holds when both agents choose partial market coverage. In the intermediate parameter range where the monopolist chooses partial market coverage while the planner serves all consumers, there exists a level of the marginal willingness to pay at which the monopolist and the planner supply the same number of varieties, although qualities as well as output levels differ across regimes. This has a relevant policy implication. Whenever $\overline{\theta} > 2 + (6\xi c)^{1/3}$, the number of varieties supplied under monopoly exceeds the socially optimal one. Hence, in this range, the introduction of an MQS able to induce the monopolist to drop some low-quality goods might be adopted by a public agency so as to enhance social welfare, provided that this does not lead the monopolist to switch to partial market coverage. This policy measure would mitigate both kinds of distortions associated with the monopolist's quality range, namely, the undersupply of quality as well as the excessive number of products. Finally, the possibility that an analogous policy may produce a welfare improvement also in the interval $\overline{\theta} \in \left(6^{1/3}(3^{2/3} + 3(\xi c)^{1/3})/3, 2 + (6\xi c)^{1/3}\right)$ can be ruled out on the following grounds. Over this range, the monopolist is pricing out of the market some low-income consumers, while supplying too many varieties. A reduction in their number would entail a reduction in output, with negative consequences on both producer and consumer surplus.

1.6 Conclusions

Using a discrete model, I have shown that we should expect a monopolist offering a vertically differentiated range of varieties of the same good to behave according to the rules found by Spence (1975). I have illustrated the case where the distribution of consumers is uniform over a given support representing marginal willingness to pay for quality. The continuous case is obtained in the limit, as the number of varieties tends to infinity. The main conclusion emerging from the continuous model, namely, that no distortion at the top should be observed in equilibrium, i.e., all consumers but that (or those) characterised by the highest valuation of quality should buy a lower quality than under social planning, is qualified by establishing whether the market is fully or only partially served. Under partial market coverage, which obtains whenever the market is sufficiently poor to induce the monopolist to price so as to exclude some individuals from consumption, qualities coincide with the socially optimal one and the misallocation of consumers across qualities is entirely due to the price mechanism distorting output. Otherwise, under full market coverage, which obtains when the monopolist cannot profit from pricing any consumer out because the market is too rich to allow it, then no distortion in the output level can be operated and the misallocation of consumers across varieties is due to the monopolist stretching the quality range below the lower bound of the socially efficient spectrum.

Then, I have endogenised the number of varieties offered by the monopolist and the planner, under the assumption that a fixed fee is associated with each variety. In this setting, I have shown that the profit incentive induces the monopolist to supply fewer varieties than the planner, when both agents cover the market only partially. The opposite holds under full market coverage, where the monopolist undersupplies quality and offers too many varieties. In such a situation, regulating the monopolist's behaviour through the adoption of an MQS may enhance social welfare.

References

1] Besanko, D., S. Donnenfeld and L. White (1987), "Monopoly and Quality Distortion: Effects and Remedies", *Quarterly Journal of Economics*, **102**, 743-68.

2] Besanko, D., S. Donnenfeld and L. White (1988), "The Multiproduct Firm, Quality Choice, and Regulation", *Journal of Industrial Economics*, **36**, 411-29.

3] Champsaur, P. and J.-C. Rochet (1989), "Multiproduct Duopolists", *Econometrica*, **57**, 533-57.
4] Cremer, H. and J.-F. Thisse (1991), "Location Models of Horizontal Differentiation: A Special Case of Vertical Differentiation Models", *Journal of Industrial Economics*, **39**, 383-90.
5] Cremer, H. and J.-F. Thisse (1994), "Commodity Taxation in a Differentiated Oligopoly", *International Economic Review*, **35**, 613-33.
6] De Meza, D. (1997), "Product Diversity under Monopoly: Two High-Quality Results", *Bulletin of Economic Research*, **49**, 169-71.
7] Donnenfeld, S. and L. White (1988), "Product Variety and the Inefficiency of Monopoly", *Economica*, **55**, 393-401.
8] Dupuit, J. (1854), *Traité théorique et pratique de la conduite et de la distribution des eaux*, Paris, De Lacroix-Cornon.
9] Ekelund, R.B. (1970), "Price Discrimination and Product Differentiation in Economic Theory: An Early Analysis", *Quarterly Journal of Economics*, **84**, 268-78.
10] Gabszewicz, J.J., A. Shaked, J. Sutton and J.-F. Thisse (1986), "Segmenting the Market: The Monopolist Optimal Product Mix", *Journal of Economic Theory*, **39**, 273-89.
11] Itoh, M. (1983), "Monopoly, Product Differentiation and Economic Welfare", *Journal of Economic Theory*, **31**, 88-104.
12] Lambertini, L. (1997), "The Multiproduct Monopolist under Vertical Differentiation: An Inductive Approach", *Recherches Economiques de Louvain*, **63**, 109-22.
13] Lambertini, L. and R. Orsini (2000), "Process and Product Innovation in a Vertically Differentiated Monopoly", *Economics Letters*, **68**, 333-37.
14] Maskin, E. and J. Riley (1984), "Monopoly with Incomplete Information", *RAND Journal of Economics*, **15**, 171-96.
15] Mussa, M., and S. Rosen (1978), "Monopoly and Product Quality", *Journal of Economic Theory*, **18**, 301-17.
16] Shaked, A. and J. Sutton (1983), "Natural Oligopolies", *Econometrica*, **51**, 1469-83.
17] Sheshinski, E. (1976), "Price, Quality and Quantity Regulation in Monopoly Situations", *Economica*, **43**, 127-37.
18] Spence, A.M. (1975), "Monopoly, Quality and Regulation", *Bell Journal of Economics*, **6**, 417-29.
19] White, L. (1977), "Market Structure and Product Varieties", *American Economic Review*, **67**, 179-82.

Chapter 2

Multiproduct monopoly with positional externalities

Luca Lambertini and Raimondello Orsini

2.1 Introduction

Most of the existing literature on vertical differentiation has focussed upon the case where the good being supplied only conveys intrinsic (hedonic) utility. However, casual observation shows that, in particular in vertically differentiated markets, purchase can be driven by social motivation rather than intrinsic characteristics. Consumption behaviours can be classified according to the distinction between material and positional goods (Hirsch, 1976; Frank, 1985a,b), or between functional and non-functional motivation (Leibenstein, 1950).[1]

Trying to achieve social distinction, individuals choose to consume goods which, either for their quality and price or for their limited supply, remain out of reach for a considerable part of consumers. In the existing literature on positional or snob goods, the external effect in consumption due to social concerns is generally modeled through a relative-consumption mechanism. The utility derived from consumption is a function of the quantity purchased relative to the average of the society or the reference group to whom the consumer compares. This approach tends to ignore that very often the social distinction accrues not from the quantity purchased, but from the quality of the chosen good. The choice can be viewed as dichotomous: some consumers

[1] The opposite situation is that where consumption generates a network externality. This case is investigated in Lambertini and Orsini (2001, 2003).

buy the positional good, some others do not. In a partial equilibrium framework, we analyse the behaviour of a monopolist selling a positional good, under two alternative assumptions. First, we consider the case where the consumers who are not served simply do not buy.[2] Second, we envisage the case where the consumers who cannot afford the positional good purchase a lower quality non-positional variety, which is bought only for its intrinsic (material or functional) characteristics.

We model the positional effect as follows. The utility that a generic consumer obtains from positional consumption is inversely related to the number of other consumers buying the same good. Therefore, the producer's choice of the output level can be compared to the provision of product quality, in that restricting output is essentially like providing consumers with a good characterised by a superior quality. Accordingly, the presence of status effects in markets where intrinsic quality is endogenous requires a re-examination of the welfare distortion associated with market power.

The theoretical framework we adopt is based upon Spence's (1975) seminal paper on vertical differentiation, which we extend by introducing a positional effect into consumer preferences. First, we summarise the analysis of the single-product case, investigated in detail in Lambertini and Orsini (2002). In this setting, it is known that, if positional effects were absent, a profit-maximising monopolist would supply the same quality as a social planner, as long as partial market coverage obtains (Spence, 1975).[3] We find that the downward quality distortion affecting the monopoly optimum is counterbalanced by the output expansion generated by an increase in the positional externality. As a consequence, the need for regulating the monopolist tends to shrink as the external effect becomes heavier.

Then, under the same assumptions on technology and consumers' distribution, we analyze the case of a two-product monopolist under full market coverage. Consumers have to choose among a high-quality (positional) good, and a low-quality (non-positional) good. The first finding is that the profit-seeking monopolist finds it always profitable to supply a non-positional low-quality good. Moreover, the only control variable being affected by the

[2]For a thorough analysis of the behaviour of a monopolist in a market where buyers' satisfaction increases in the number of consumers excluded from purchase, see Basu (1987), suggesting that the positional concern may be traced back to either quality signalling by the firm or status seeking by consumers.

[3]See also Sheshinski (1976). Spence's analysis has been extended to the multiproduct case by Mussa and Rosen (1978), Itoh (1983) and Champsaur and Rochet (1989), where a continuum of qualities is considered. Gabszewicz *et al.* (1986) characterise the optimal product mix of the monopolist when technology involves fixed costs. The provision of high quality under monopoly is analysed in de Meza (1997).

positional effect is the price of the high-quality good. Qualities and outputs are the same as in the standard model without positional concerns.

The social incentive towards the introduction of a non-positional variety exists if the positional concern is sufficiently high (or, equivalently, if the market is sufficiently poor). As in the case of the profit-seeking monopolist, the price of the positional good is used to fully internalise the externality, while qualities and outputs are the same as in the standard model without positional concerns. Finally, under full market coverage, the welfare loss associated with the monopoly equilibrium is unrelated to the extent of the positional externality.

In both regimes, the introduction of a non-positional good entails an enhancement in the quality level of the positional one, as compared to the single-product case.

The chapter is organized as follows. Section 2.2 presents the single-good model, showing the monopoly equilibrium and the social optimum, which are then comparatively evaluated. Section 2.3 presents the two-product monopoly, whose welfare performance is assessed against the behaviour of a social planner. Section 2.4 contains concluding remarks.

2.2 Single-good monopoly

To begin with, we briefly summarise the model investigated in Lambertini and Orsini (2002). A monopolist operates in a market for a good whose utility depends on intrinsic characteristics, which are represented by quality q, as well as the social status enjoyed by consumers through its purchase. Social distinction is driven by the fact that the market is only partially covered, i.e. there are some individuals who do not consume. Therefore, preferences exhibit a positional externality, whose amount depends inversely on market demand x. Except for the externality, the preference structure is the same as in Mussa and Rosen (1978). That is, consumers are indexed by parameter θ, measuring the individual marginal willingness to pay for quality. Consumers are uniformly distributed over the interval $[\bar{\theta}-1, \bar{\theta}]$, with $\bar{\theta} \geq 1$. The total population is normalised to 1. Each consumer buys one unit of the good maximizing the net surplus he obtains, provided that it is non-negative:

$$U = \theta q + \alpha(1 - x) - p \tag{2.1}$$

where p is the price charged by the monopolist, while α (which is assumed to be the same across agents) is a positive parameter representing the weight of the positional externality in the utility function, and x is the market demand for the good. Hence, $1 - x$ is the number of consumers who are unable to buy. This implies that, *ceteris paribus*, an additional purchase by a consumer previously priced out of the market produces a negative externality on the group of individuals who were already being served. It is worth noting that, under full market coverage, the positional externality disappears. This accounts for the fact that, when all consumers buy, the purchase of the good does not yield any social distinction.

Market demand is $x = \bar{\theta} - \widehat{\theta}$, where $\widehat{\theta}$ identifies the marginal willingness to pay for quality of the marginal consumer, i.e., that individual who is indifferent between buying or not. For this consumer, the following equality must hold:

$$U\left(\widehat{\theta}\right) = \widehat{\theta}q + \alpha\left(1 - \bar{\theta} + \widehat{\theta}\right) - p = 0\,, \tag{2.2}$$

yielding $\widehat{\theta}(\alpha, p, q) = (p + \alpha\bar{\theta} - \alpha)/(q + \alpha)$. Hence, under partial market coverage and full market coverage, respectively, market demand is

$$\begin{gathered} x = \bar{\theta} - \widehat{\theta}(\alpha, p, q) = \frac{\bar{\theta}q - p + \alpha}{q + \alpha} \text{ for all } \{p, q, \alpha\} \\ \text{such that } \hat{\theta}(\alpha, p, q) \in (\bar{\theta} - 1, \bar{\theta}] \quad (pmc); \end{gathered} \tag{2.3}$$

$$x = 1 \text{ for all } \{p, q, \alpha\} \text{ such that } \hat{\theta}(\alpha, p, q) \leq \bar{\theta} - 1 \quad (fmc). \tag{2.4}$$

When $\max\left\{\bar{\theta} - 1, \widehat{\theta}(\alpha, p, q)\right\} = \bar{\theta} - 1$, full market coverage ($fmc$) obtains, i.e., $x = 1$, and the positional externality disappears. On the supply side, production takes place at a total cost equal to $C = q^2x$. That is, we assume that there are constant returns to scale and unit production costs $c(q)$ are convex in quality. The profit function is then

$$\Pi^M = \left(p - q^2\right) \cdot x\,, \tag{2.5}$$

where superscript M stands for *monopoly*. Except for the presence of positional effects, this setup replicates the model introduced by Spence (1975).[4] He proves that, for a given output level, if consumers are uniformly distributed and the market is covered only partially, the monopolist supplies the socially optimal quality. In the remainder of the section, we will show that this is not the case when a positional concern is present.

2.2.1 Profit maximization

Since we are focusing on the effects of social distinction on market behaviour, we confine our analysis to the case of partial market coverage, i.e., $\hat{\theta}(\alpha, p, q) \in (\bar{\theta} - 1, \bar{\theta}]$. Monopoly profits are given by:

[4] Spence (1975), however, does not specify the utility function in the form that is later introduced by Mussa and Rosen (1978).

$$\Pi^M = (p - q^2) \cdot \left[\frac{\bar{\theta} q - p + \alpha}{q + \alpha} \right] . \tag{2.6}$$

This expression has to be maximized with respect to the two choice variables: price and quality.[5] The first order conditions (FOCs) with respect to price and quality are:

$$\frac{\partial \Pi^M}{\partial p} = \frac{\alpha + \bar{\theta} q - 2p + q^2}{q + \alpha} = 0 ; \tag{2.7}$$

$$\frac{\partial \Pi^M}{\partial q} = \frac{-2\bar{\theta} q^3 + q^2(p - \alpha - 3\alpha\bar{\theta}) + 2\alpha\bar{\theta}(p - \alpha) + p^2 + \alpha\bar{\theta} p - \alpha p}{(q + \alpha)^2} = 0 . \tag{2.8}$$

Solving the system (2.7-2.8) yields:

$$p^M = \frac{3\alpha + 4\alpha^2 - 2\alpha\bar{\theta} + \bar{\theta}^2 + (\bar{\theta} - \alpha) \cdot k}{9} ; \tag{2.9}$$

$$q^M = \frac{\bar{\theta} - 4\alpha + k}{6} , \tag{2.10}$$

where $k = \sqrt{\bar{\theta}^2 + 16\alpha\bar{\theta} + 16\alpha^2 - 12\alpha}$. Observe that, positional concern being absent ($\alpha = 0$), $q^M = \bar{\theta}/3$ and $p^M = 2\bar{\theta}^2/9$ (see Lambertini, 1997). Market demand in equilibrium amounts to:

$$x^M = \frac{\bar{\theta}^2 - 8\alpha\bar{\theta} + 12\alpha - 8\alpha^2 + (2\alpha + \bar{\theta})k}{3(\bar{\theta} + 2\alpha + k)} . \tag{2.11}$$

Consumer surplus is

$$CS^M = \int_{\widehat{\theta}}^{\bar{\theta}} [\theta q^M + \alpha(1 - x^M) - p^M] d\theta . \tag{2.12}$$

Social welfare corresponds to $SW^M = CS^M + \Pi^M$. The comparative statics properties of the positional externality on price, quality and output are summarised by

Lemma 2.1 *Monopoly price is always increasing in* α. *Moreover,* $x^M < 1$ *for all* $\bar{\theta} \in \left[1, 2 + \sqrt{4\alpha + 1}\right)$. *The following properties can be shown to hold:*

[5] The choice of maximising profits w.r.t. price or quantity is obviously irrelevant, with the same results obtaining in the two cases.

(i) $\lim_{\alpha\to 0} x^M = \overline{\theta}/3$; $\lim_{\alpha\to\infty} x^M = 1/2$;

(ii) for all $\overline{\theta} \in [1, 3/2)$, $\partial x^M/\partial\alpha > 0$ and $\partial q^M/\partial\alpha < 0$;

(iii) for all $\overline{\theta} \in \left(3/2, 2+\sqrt{4\alpha+1}\right)$, $\partial x^M/\partial\alpha < 0$ and $\partial q^M/\partial\alpha > 0$;

(iv) for $\overline{\theta} = 3/2$, $\partial x^M/\partial\alpha = \partial q^M/\partial\alpha = 0$, *with* $x^M = q^M = 1/2$.

To verify the behaviour of x^M as α changes, it suffices to examine:

$$\frac{\partial x^M}{\partial\alpha} = 4[30\alpha^2 - 18\alpha - 16\alpha^3 + 30\alpha\overline{\theta} - 24\alpha^2\overline{\theta} + 3\overline{\theta}^2 - 12\alpha\overline{\theta}^2 - 2\overline{\theta}^3 +$$

$$+k(4\alpha^2 - 6\alpha + 3\overline{\theta} + 4\alpha\overline{\theta} - 2\overline{\theta}^2)]/[3k(2\alpha + \overline{\theta} + k)]^2, \qquad (2.13)$$

whose sign changes along $\overline{\theta} = 3/2$. The behaviour of market demand x^M as α varies, for given levels of $\overline{\theta}$, is represented in figure 2.1, which obtains by drawing x^M for three specific values of $\overline{\theta} \in \left[1, 2+\sqrt{4\alpha+1}\right)$.

Figure 2.1 : Monopoly output

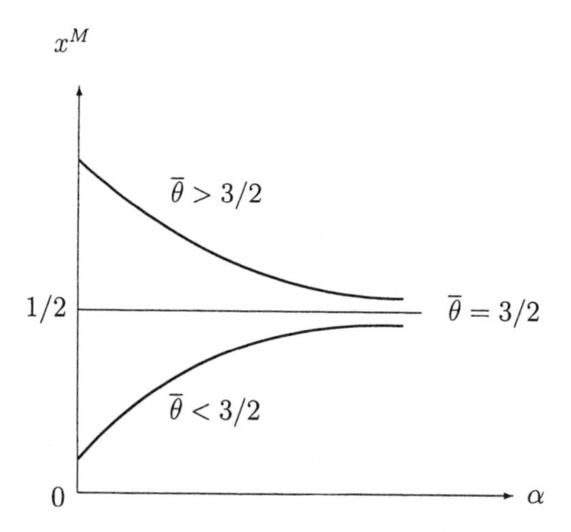

Notice that, depending on the level of $\overline{\theta}$, equilibrium quality levels tend to diverge as the amount of positional externality α increases. This phenomenon is described in figure 2.2.

Figure 2.2 : Monopoly quality

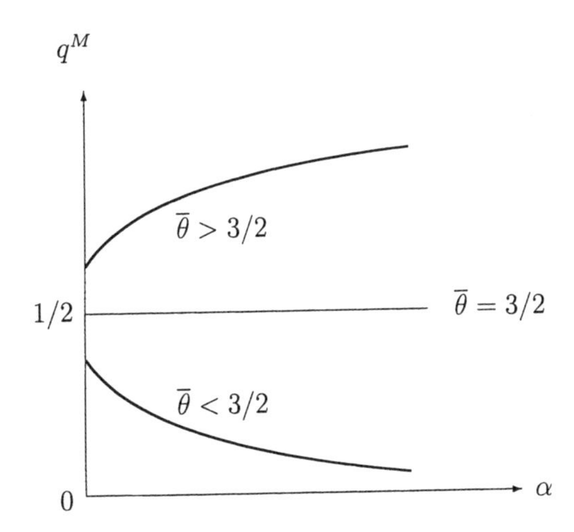

The main properties of the monopolist's optimal behavior can be summarised in the following terms.

A] Suppose $\overline{\theta} = 3/2$. If so, then $q^M = x^M = 1/2$, and the net surplus of the marginal consumer at $\widehat{\theta}$ is $U_{\widehat{\theta}} = \widehat{\theta} q^M - p^M + \alpha(1 - x^M) = (1 - 2p^M + \alpha)/2 = 0$ at $p^M = (1 + \alpha)/2$. This entails that, since p^M increases in α as fast as the positional effect $\alpha(1 - x^M)$, along $\overline{\theta} = 3/2$ the identity of the marginal consumer, and therefore the size of demand, is independent of α.

B] Suppose $\overline{\theta} \in \left(3/2, 2 + \sqrt{4\alpha + 1}\right)$. In this range, equilibrium demand is larger. However, as α increases, the monopolist supplies a better quality serving fewer consumers. When the marginal willingness to pay for quality is high, the positional effect leads to an *elitarian equilibrium*, with few buyers of an expensive high-quality good.

C] Suppose $\overline{\theta} \in [1, 3/2)$. In this parameter range, demand is low. Yet, as α increases, the monopolist reduces the quality level, consequently expanding market coverage. When the marginal willingness to pay for quality is low, the presence of a positional effect is simply exploited by the monopolist so as to increase price in spite of the reduction in quality and the expansion in demand.

As to social welfare and its components, the following holds:

Proposition 2.1 *Monopoly profits are always increasing in the extent of the positional externality, while social welfare is increasing (decreasing) in α for sufficiently low (high) levels of $\overline{\theta}$.*

Proof. See Lambertini and Orsini (2002, pp. 157-8). ■

The critical level of $\overline{\theta}$ around which the sign of $\partial SW^M/\partial\alpha$ changes is $\overline{\theta} \cong 3.80\left(0.34+\sqrt{0.06+0.26\alpha}\right)$. The non-monotonicity of social welfare w.r.t. positional effects is broadly determined by the analogous non-monotonicity of monopoly output.

2.2.2 Welfare maximization

Now move on to describe the behaviour of a benevolent planner choosing price and quality so as to maximise social welfare. The first order condition for welfare maximisation w.r.t. price is:

$$\frac{\partial SW^{SP}}{\partial p} = \frac{\alpha^2+\alpha\overline{\theta}q-2\alpha p-pq+\alpha q^2+q^3}{(q+\alpha)^2} = 0\,, \tag{2.14}$$

where superscript SP stands for *social planning*. This yields $p^{SP} = (\alpha^2 + \alpha\overline{\theta}q + \alpha q^2 + q^3)/(2\alpha + q)$. Then, solving $\partial SW^{SP}/\partial q = 0$ yields optimal quality:

$$q^{SP} = \frac{\overline{\theta}-8\alpha+h}{6}\,, \tag{2.15}$$

where $h = \sqrt{\overline{\theta}^2+32\alpha\overline{\theta}-12\alpha+64\alpha^2}$. It is easy to verify that $\partial q^{SP}/\partial\alpha > 0$ always, and $\lim_{\alpha\to\infty} q^{SP} = (4\overline{\theta}-1)/8$. The associated price rewrites as

$$p^{SP} = \frac{216\alpha^2+\left[\alpha(16\alpha-10h+26\overline{\theta})+(\overline{\theta}+h)^2\right](\overline{\theta}-8\alpha+h)}{36(\overline{\theta}+4\alpha+h)}\,. \tag{2.16}$$

It can be verified that $\partial p^{SP}/\partial\alpha > 0$. The equilibrium output is

$$x^{SP} = \frac{2\left[\overline{\theta}\left(\overline{\theta}-16\alpha\right)+4\alpha\left(3-8\alpha\right)+(4\alpha+\overline{\theta})h\right]}{3(4\alpha+\overline{\theta}+h)}\,, \tag{2.17}$$

with full market coverage obtaining at $\alpha = \alpha' = (4\overline{\theta}^2-8\overline{\theta}+3)/16$, which is larger than $\widehat{\alpha}$ for all $\alpha > 0$ and $\overline{\theta} > 1$. Moreover, $\alpha' > 0$ for all $\overline{\theta} > 3/2$, and $\partial x^{SP}/\partial\alpha < 0$ always.

This proves the following lemma:

Lemma 2.2 *At the social optimum, there emerges an elitarian equilibrium where $\partial x^{SP}/\partial\alpha < 0$ always. Moreover,* *(i)* $x^{SP} > x^M$ *for all levels of* $\overline{\theta}$ *and* α *such that* $x^M < 1$;

(ii) $\partial SW^{SP}/\partial\alpha > 0$ *for all levels of* $\overline{\theta}$ *and* α *such that* $x^{SP} < 1$.

Now we briefly assess the relative performances of the profit-seeking monopolist and the social planner offering a single product each. This can be done by comparing quality levels q^M and q^{SP} as well as social welfare levels SW^M and SW^{SP}, yielding the following:

Proposition 2.2 *For all* $\{\alpha, \overline{\theta}\}$, $q^M < q^{SP}$, *i.e., the monopolist undersupplies product quality. However,* $\partial\left(SW^{SP} - SW^M\right)/\partial\alpha < 0$ *for all* $\{\alpha, \overline{\theta}\}$ *such that partial market coverage obtains under both regimes.*

The above result derives from the fact that the quality distortion affecting the monopoly optimum is counterbalanced by the output expansion generated by an increase in the positional externality. Accordingly, the need for regulating the monopolist tends to shrink as the external effect becomes heavier.

2.3 The multiproduct monopolist with full market coverage

Here, we investigate how the picture changes as soon as those consumers who are unable to purchase a positional good have access to some lower-quality (cheaper) alternative. Consider the case of a monopoly supplying two products characterised by quality levels $q_H \geq q_L \geq 0$.[6] In particular, we focus upon the case where the positional externality is associated with q_H, and all consumers who do not purchase the positional high-quality good buy an alternative low-quality good which does not confer any *status*. Hence, we confine our attention to the full market coverage setting. As an illustration, suppose we are describing the car market. The population of potential buyers is distributed over $\left[\overline{\theta} - 1, \overline{\theta}\right]$. The high-quality positional good could be a flagship sportscar like Porsche or Ferrari, conferring social distinction. Such a car is accessible to rich consumers only, while others may buy anything, like Fiat Punto or VW Polo, which can be assumed to be non-positional. All consumers in $\left[0, \overline{\theta} - 1\right]$ use either buses or the tube.

[6] In an alternative setting, Bagwell and Bernheim (1996) describe a market for *conspicuous goods* where products are homogeneous, but a *griffe* adds a status signal to the intrinsic use of a product. This entails pricing above marginal cost by the firm producing the conspicuous good.

The net surplus of a consumer who buys the high quality good is

$$U_H = \theta q_H + \alpha(1 - x_H) - p_H \ , \tag{2.18}$$

where $1 - x_H = x_L$, while that of a consumer buying the low-quality good is $U_L = \theta q_L - p_L$. Along the support $\left[\overline{\theta} - 1, \overline{\theta}\right]$, define as θ' the location of the consumer who is indifferent between q_H and q_L at generic prices $\{p_H, p_L\}$. Therefore, $1 - x_H = x_L = 1 - \overline{\theta} + \theta'$ and θ' obtains from the solution of $U_H = U_L$:

$$\theta' = \frac{\alpha(\overline{\theta} - 1) + p_H - p_L}{\alpha + q_H - q_L} \ . \tag{2.19}$$

Consumer surplus in the two market segments is, respectively:

$$CS_H = \int_{\theta'}^{\overline{\theta}} U_H d\theta \ ; \ CS_L = \int_{\overline{\theta}-1}^{\theta'} U_L d\theta \ . \tag{2.20}$$

The monopolist's profit function is

$$\Pi^M = \sum_i (p_i - q_i^2) x_i \ , \ i = H, L \ . \tag{2.21}$$

Therefore, social welfare amounts to

$$SW = CS_H + CS_L + \Pi^M \ . \tag{2.22}$$

2.3.1 Profit-maximising monopoly

The monopolist chooses prices and qualities to maximise the following profit function:

$$\Pi^M = \left(p_H - q_H^2\right)\left(\overline{\theta} - \theta'\right) + \left(p_L - q_L^2\right)\left(\theta' - \overline{\theta} + 1\right) \tag{2.23}$$

We proceed as follows. First, we solve the monopolist's pricing problem, which consists in extracting all the surplus from the poorest consumer through

$$p_L^M = \left(\overline{\theta} - 1\right) q_L \tag{2.24}$$

and solving the FOC w.r.t. p_H :

$$\frac{\partial \Pi^M}{\partial p_H} = \frac{\alpha + (q_H - q_L)\left(\overline{\theta} + q_H + q_L\right) - 2(p_H - p_L)}{\alpha + q_H - q_L} \tag{2.25}$$

which yields

$$p_H^M = \frac{\alpha + (q_H + q_L)\left(\overline{\theta} + q_H - q_L\right) - 2q_L}{2} \ . \tag{2.26}$$

Then, equilibrium qualities are the solution to the system:

$$\frac{\partial \Pi^M}{\partial q_H} = 0 \; ; \; \frac{\partial \Pi^M}{\partial q_L} = 0 \, . \tag{2.27}$$

The system (2.27) has nine critical points identifying quality pairs belonging to $\mathbb{R}$. Among them, we select the only pair of qualities which (i) satisfy second order conditions and (ii) when $\alpha = 0$, coincide with the equilibrium qualities observed in the well known model without externality (see Lambertini, 1997). This produces the following:

Lemma 2.3 *Optimal monopoly qualities are independent of the positional externality:*

$$q_H^M = \frac{2\overline{\theta} - 1}{4} \; ; \; q_L^M = \frac{2\overline{\theta} - 3}{4} \, . \tag{2.28}$$

Comparing q_H^M in (2.28) with (2.10) we have that

$$\lim_{\alpha \to \infty} \left(q_H^M - q^M \right) = 0 \text{ for all } \overline{\theta} \; ; \tag{2.29}$$

in particular, in the *elitarian equilibrium* $\left(\overline{\theta} \in \left(3/2, 2 + \sqrt{4\alpha + 1}\right)\right)$, we observe:

$$q_H^M > q^M \; ; \; \frac{\partial \left(q_H^M - q^M \right)}{\partial \alpha} < 0 \, . \tag{2.30}$$

That is, the introduction of a non-status variety induces the monopolist to raise the quality level of the positional good, as compared to the case where the latter is the only variety supplied to the market.[7]

Plugging equilibrium qualities (2.28) into prices, we obtain:

$$p_H^M = \frac{\overline{\theta}\left(2\overline{\theta} - 3\right) + 2}{4} + \frac{\alpha}{2} \; ; \; p_L^M = \frac{\left(\overline{\theta} - 1\right)\left(2\overline{\theta} - 3\right)}{4} \tag{2.31}$$

and the associated equilibrium demands are $x_H^M = x_L^M = 1/2$. This, in turn, entails a further conclusion:

Lemma 2.4 *The only control being affected by the positional externality is the price of the high-quality positional good.*

[7] When $\overline{\theta} \in [1, 3/2]$, the following holds:

$$\overline{\theta} < \frac{3}{2} \Rightarrow q_H^M < q^M$$

$$\overline{\theta} = \frac{3}{2} \Rightarrow q_H^M = q^M = \frac{1}{2}$$

The pricing policy of the monopolist is designed so as to fully internalise the positional effect, so as to keep the distribution of consumers unchanged with respect to the case without positional externality. This can be quickly verified by checking that $\partial p_H^M/\partial\alpha = 1/2 = x_H^M$. Ultimately, this implies that consumer surplus is unaffected by positional concerns at the monopoly equilibrium. Equilibrium profits and welfare amount to:

$$\Pi^M(2) = \frac{4\overline{\theta}\left(\overline{\theta}-2\right)+5}{16} + \frac{\alpha}{4}\,;\; SW^M(2) = \frac{\overline{\theta}\left(\overline{\theta}-1\right)+\alpha}{4}\,. \tag{2.32}$$

Numerical calculations show that the following inequality:

$$\Pi^M(2) - \Pi^M > 0 \tag{2.33}$$

holds for all admissible values of $\left\{\alpha, \overline{\theta}\right\}$. Therefore, the monopolist always finds it profitable to enlarge the product spectrum to include a non-positional variety.

2.3.2 Social planning

A benevolent social planner chooses prices and qualities to maximise welfare (2.22). The price of the low-quality good is $p_L^{SP} \in [q_L^2, (\overline{\theta}-1)q_L]$,[8] while the price of the high-quality good is the solution to the first order condition $\partial SW/\partial p_H = 0$:

$$p_H^{SP} = p_L^{SP} + \frac{\alpha(\alpha+\overline{\theta}q_H+q_H^2-\overline{\theta}q_L-q_L^2)+(q_H^2-q_L^2)(q_H-q_L)}{q_H-q_L+2\alpha}\,. \tag{2.34}$$

This entails that prices are not linearly independent, and there exist infinitely many price pairs $\left\{p_H^{SP}, p_L^{SP}\right\}$ allowing the planner to maximise social welfare.[9] The price of the high-quality good can be rewritten as follows:

$$p_H^{SP} = p_L^{SP} + q_H^2 - q_L^2 + \alpha x_L \tag{2.35}$$

which obviously coincides, if $\alpha = 0$, with the socially optimal price in absence of the positional externality (see Lambertini and Mosca, 1999). For any quality pair, the planner extracts from the status-seeking consumers a tax revenue of the same size as the positional externality.

[8] We impose that the break even condition on the low-quality good be satisfied. However, in a general equilibrium analysis, the planner could also price the low-quality good below its marginal cost.

[9] Observe that this stems from the assumption of full market coverage, which implies that the size of the social surplus is given (see Lambertini and Mosca, 1999). Under partial market coverage, the planner would price the low-quality good at marginal cost.

The system $\{\partial SW/\partial q_H = 0; \partial SW/\partial q_L = 0\}$ of first order conditions w.r.t. qualities has five critical points, out of which

$$q_H^{SP} = \frac{4\overline{\theta} - 1}{8};\ q_L^{SP} = \frac{4\overline{\theta} - 3}{8} \tag{2.36}$$

is the only pair satisfying second order conditions. Notice that q_H^{SP} and q_L^{SP} do not depend on α. In particular, they coincide with the socially optimal qualities associated with the maximisation of social welfare in a standard model without positional externality (see Lambertini, 1997). Moreover, a straightforward comparison between (2.36) and (2.28) reveals that the monopolist's incentive to induce self-selection across consumers causes an increasing distortion in the quality level as one moves downward along the quality spectrum. Finally, notice that $q_H^{SP} > q^{SP}$ for all $\alpha \in [0, \infty)$ (cf. expression (2.15) above), that is, the quality characterising the positional good under social planning becomes higher when the planner also supplies a low-quality alternative.

Social welfare amounts to $SW^{SP}(2) = [16(\overline{\theta}^2 - \overline{\theta} + \alpha) + 5]/64$, where the number in brackets indicates that the planner supplies two varieties. Equilibrium prices are $p_H^{SP} \in [(16\overline{\theta}^2 - 8\overline{\theta} + 1)/64 + \alpha/2, (4\overline{\theta}^2 - 5\overline{\theta} + 2)/8 + \alpha/2]$ and $p_L^{SP} \in [(4\overline{\theta} - 3)^2/64, (\overline{\theta} - 1)(4\overline{\theta} - 3)/8]$, while output levels are $x_H^{SP} = x_L^{SP} = 1/2$. Together with Lemma 4, this discussion proves the following:

Proposition 2.3 *Both the planner and the monopolist increase the price of the positional good, so as to fully internalise the positional externality. Under both regimes, the optimal allocation of consumers obtains when the richer (poorer) half of the population buys the high (low) quality good.*

Comparing the output levels $x_H^{SP} = x_L^{SP} = 1/2$ with the quantity produced by the planner operating with a single product (expression (2.17)), an additional result emerges:

Corollary 2.1 $x_H^{SP} = 1/2 < x^{SP}$ *for all admissible values of* α *and* $\overline{\theta}$.

The above corollary states that the introduction of a non-positional good in the lower segment of the quality spectrum entails a reduction in the provision of the positional variety.

In the light of the foregoing analysis, a further issue arises, namely, whether the conventional result, according to which the welfare performance of the planner would improve as the number of varieties increases (see Mussa and Rosen, 1978; Champsaur and Rochet, 1989; Lambertini, 1997), holds true in the present setting as well. In order to answer this question, it suffices to compare SW^{SP} against $SW^{SP}(2)$. Define $\Delta SP = SW^{SP} - SW^{SP}(2)$, which is plotted over $\{\alpha, \overline{\theta}\}$ in figure 2.3.

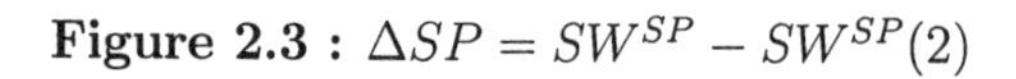

Figure 2.3 : $\Delta SP = SW^{SP} - SW^{SP}(2)$

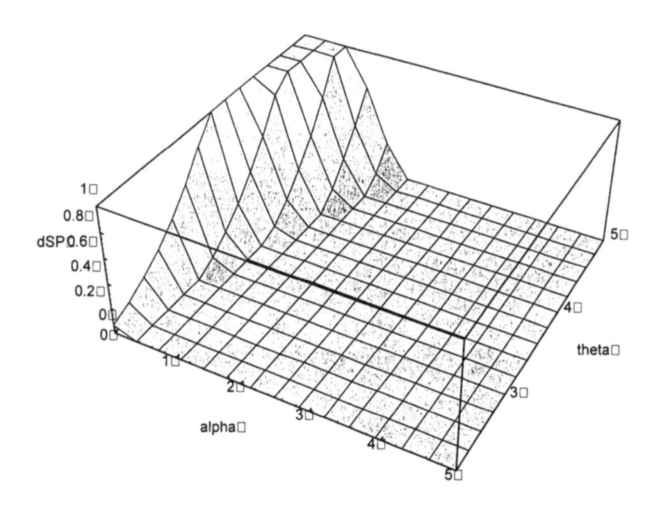

Obviously, the viable region wherein it makes sense to carry out such comparison is $\alpha \in [0, \alpha')$. Figure 3 highlights that there is a wide subset of this region of the relevant parameters $\{\alpha, \overline{\theta}\}$ where it is not convenient for the planner to introduce a non-positional low-quality good.[10] The explanation of this phenomenon is to be looked for in the behaviour of qualities and prices as the planner switches from one to two varieties. The decrease in the output and the associated increase in the quality level of the positional good seem to point to a welfare improvement for those consumers who purchase the positional variety independently of the number of products supplied by the planner. The welfare effect should be positive also for those consumers located in the lower part of the range of θ, who are in a position to buy only when two products are available. However, consider that $\Delta SP = SW^{SP} - SW^{SP}(2) \geq 0$ for low values of α. In this region, qualities are approximately linear in $\overline{\theta}$, while marginal production costs as well as prices are approximately quadratic in $\overline{\theta}$, so that gross consumer surplus θq

[10] To emphasise this point, in figure 6 we confine the plot to the positive range of ΔSP. This entails that the flat region of the plot denotes all the pairs $\{\alpha, \overline{\theta}\}$ for which $\Delta SP \leq 0$.

increases less rapidly than marginal costs. Besides, the lower is α, the larger becomes the difference $q_H^{SP} - q^{SP}$, making heavier the cost burden associated with expanding the product range.

2.3.3 Monopoly vs social planning

The welfare comparison between social planning and monopoly with two varieties reduces to the straightforward evaluation of the following:

$$\Delta SW = SW^{SP}(2) - SW^{M}(2) = \frac{5}{64} \tag{2.37}$$

which proves:

Proposition 2.4 *The incentive for a benevolent policy maker to regulate the behaviour of the profit-maximising monopolist is independent of the extent of the positional externality.*

This contrasts with the discussion carried out in section 2.2 concerning the single-product monopolist, where the welfare distortion shrinks as α grows larger. This is due to the fact that, if only one variety is supplied, partial market coverage is needed for the positional externality to be effective. This implies that, lacking a non-positional substitute, the pricing behaviour affects the size of equilibrium demand and therefore it is not optimal to set the price so as to fully internalise the externality.

2.4 Concluding remarks

We have modeled the role of positional effects in a market for vertically differentiated goods. We have considered two alternative settings, namely, a single-product monopoly with partial market coverage (as in Lambertini and Orsini, 2002) and a two-product monopoly with full market coverage. In the second case, the positional externality is associated with the high-quality good, while the low-quality variety is non-positional. Then, we have evaluated the welfare performance of both settings against the behaviour of a social planner operating with the same number of varieties.

The welfare distortion caused by monopoly power is always non-increasing in the extent of the positional externality, irrespective of the number of products being supplied. However, the policy implications are different in the two cases. On the one hand, under partial market coverage, the welfare loss due to monopoly power is strictly decreasing in the extent of the positional externality, so that the scope for regulation shrinks as the positional effect becomes more relevant. On the other hand, under full market coverage, the welfare loss is independent of the positional concern.

As to the bearings of positional effects upon quality levels, our conclusions are conditional upon the extent of market coverage. When only the positional good is supplied, the equilibrium quality increases (respectively, decreases) in the positional effect if the market is sufficiently rich (poor). When a non-positional good is also supplied so as to cover the whole market, product qualities are unaffected by the positional externality. In this case, under both monopoly and social planning, the optimal allocation of consumers across goods obtains by fully internalising the positional effect through a tax applied to the high-quality product. These two factors cause the welfare distortion to be independent of the positional effect.

Finally, we have shown that the conventional wisdom, establishing that the planner's welfare performance increases in the number of varieties being supplied, may not hold in the presence of a positional concern.

References

1] Bagwell, L.S., and B.D. Bernheim (1996), "Veblen Effects in a Theory of Conspicuous Consumption", *American Economic Review*, **86**, 349-73.
2] Basu, K. (1987), "Monopoly, Quality Uncertainty and 'Status' Goods", *International Journal of Industrial Organization*, **5**, 435-46.
3] Champsaur, P., and J.-C. Rochet (1989), "Multiproduct Duopolists", *Econometrica*, **57**, 533-57.
4] de Meza, D. (1997), "Product Diversity under Monopoly: Two High Quality Results", *Bulletin of Economic Research*, **49**, 169-71.
5] Frank, R.H. (1985a), "The Demand for Unobservable and Other Nonpositional Goods", *American Economic Review*, **75**, 101-16.
6] Frank, R.H. (1985b), *Choosing the Right Pond: Human Behavior and the Quest for Status*, Oxford, Oxford University Press.
7] Gabszewicz, J.J., A. Shaked, J. Sutton and J.-F. Thisse (1986), "Segmenting the Market: The Monopolist Optimal Product Mix", *Journal of Economic Theory*, **39**, 273-89.
8] Hirsch, F. (1976), *Social Limits to Growth*, London, Routledge.
9] Ireland, N.J. (1994), "On Limiting the Market for Status Signals", *Journal of Public Economics*, **53**, 91-110.
10] Itoh, M. (1983), "Monopoly, Product Differentiation and Economic Welfare", *Journal of Economic Theory*, **31**, 88-104.

11] Lambertini, L. (1997), "The Multiproduct Monopolist under Vertical Differentiation: An Inductive Approach", *Recherches Economiques de Louvain*, **63**, 109-122.
12] Lambertini, L. and M. Mosca (1999), "On the Regulation of a Vertically Differentiated Market", *Australian Economic Papers*, **38**, 354-366.
13] Lambertini, L. and R. Orsini (2001), "Network Externalities and the Overprovision of Quality by a Monopolist", *Southern Economic Journal*, **67**, 969-82.
14] Lambertini, L. and R. Orsini (2002), "Vertically Differentiated Monopoly with a Positional Good", *Australian Economic Papers*, **41**, 151-63.
15] Lambertini, L. and R. Orsini (2003), "Monopoly, Quality, and Network Externalities", *Keio Economic Studies*, **40**, 1-16.
16] Leibenstein, H. (1950), "Bandwagon, Snob, and Veblen Effects in the Theory of Consumers' Demand", *Quarterly Journal of Economics*, **64**, 183-207. Reprinted in H. Leibenstein (1976), *Beyond Economic Man*, Cambridge, Mass., Harvard University Press, 48-67.
17] Mussa, M., and S. Rosen (1978), "Monopoly and Product Quality", *Journal of Economic Theory*, **18**, 301-17.
18] Sheshinski, E. (1976), "Price, Quality and Quantity Regulation in Monopoly Situations", *Economica*, **43**, 127-37.
19] Spence, A.M. (1975), "Monopoly, Quality and Regulation", *Bell Journal of Economics*, **6**, 417-29.

Chapter 3

Labour-managed multiproduct monopoly

Chiara Celada and Luca Lambertini

3.1 Introduction

The existing literature on the supply of product quality in monopoly markets mainly dwells upon the comparison between the behaviour of a profit-seeking monopolist and the social optimum (see Spence, 1975; Sheshinski, 1976; Mussa and Rosen, 1978; Itoh, 1983; Maskin and Riley, 1984; Gabszewicz *et al.*, 1986; Champsaur and Rochet, 1989; Lambertini, 1997a, *inter alia*). To the best of our knowledge, the only contribution dealing with a labour-managed (LM) monopoly is Lambertini (1997b).

In a well known paper, Spence (1975) has shown that, for any given output level, the profit-seeking monopolist will over- or undersupply product quality as compared to the social optimum, depending upon whether the marginal consumer is richer or poorer than the average one. Here, we want to extend Spence's analysis to account for the behaviour of the LM monopolist. First we adopt the same general approach as in Spence, to show that the LM firm may distort quality in either direction, depending upon the features of the income distribution characterising the population of consumers. Then, we illustrate the details of a commonly used specification of the model, where the distribution of income is uniform and production involves variable costs which are assumed to be quadratic in the quality level. In this setting, we prove that the LM monopolist always undersupplies quality as compared to both the profit-seeking monopolist and the benevolent planner.

3.2 The setup

The reference model is borrowed from Spence (1975). A single-product firm operates in a market where the inverse demand function is $P(x,q)$, with x and q denoting, respectively, the output level and the quality level. The properties of the demand functions are summarised by the following derivatives:

$$\frac{\partial P(x,q)}{\partial x} < 0\,; \; \frac{\partial P(x,q)}{\partial q} > 0\,. \tag{3.1}$$

The firm bears total costs $C = c(x,q) + F$, where

$$\frac{\partial c(x,q)}{\partial x} > 0\,; \; \frac{\partial c(x,q)}{\partial q} > 0\,, \tag{3.2}$$

and $F > 0$ is a fixed cost. As in Lambertini (1997b), the variable cost component is assumed to correspond to the cost of the labour input (or membership) L, $wL \equiv c(x,q)$, and the labour market wage w is normalised to one in order to simplify calculations.

The LM monopolist has to choose x and q so as to maximise value added per worker:

$$V(x,q) \equiv \frac{x \cdot P(x,q) - F}{L} = \frac{x \cdot P(x,q) - F}{c(x,q)} \tag{3.3}$$

and, of course, it must be that $V(x,q) > 1$ in order for the LM firm to attract members.

In the remainder, we shall evaluate the behaviour of an LM monopolist using maximand (3.3) against that of (i) a profit-seeking monopolist, whose objective function is:

$$\pi(x,q) = x \cdot P(x,q) - c(x,q) - F \tag{3.4}$$

and (ii) a benevolent social planner maximising welfare, defined as the sum of profits and consumer surplus:

$$SW(x.q) = \pi(x,q) + CS(x,q) \tag{3.5}$$

where

$$CS(x,q) = \int_0^x P(z,q)\,dz - x \cdot P(x,q)\,. \tag{3.6}$$

As in Spence (1975), we will carry out a comparative evaluation of the incentive to supply product quality in the three alternative regimes, *for a given output level.*

3.3 The LM monopoly optimum

The following partial derivative captures the incentive of the LM firm to supply product quality:

$$\frac{\partial V(x,q)}{\partial q} = \frac{x \cdot c(x,q) \cdot \partial P(x,q)/\partial q - [x \cdot P(x,q) - F] \cdot \partial c(x,q)/\partial q}{[c(x,q)]^2} = 0 \tag{3.7}$$

In equilibrium, the optimal quality is implicitly determined by:

$$\frac{\partial c(x,q)}{\partial q} = \frac{x \cdot c(x,q) \cdot \partial P(x,q)/\partial q}{x \cdot P(x,q) - F} \tag{3.8}$$

with the r.h.s. of (3.8) being obviously positive. To begin with, the above expression can be compared with the optimality condition for a profit-seeking monopolist:

$$\frac{\partial \pi(x,q)}{\partial q} = 0 \Leftrightarrow \frac{\partial c(x,q)}{\partial q} = x \cdot \frac{\partial P(x,q)}{\partial q}. \tag{3.9}$$

This yields the following result:

Proposition 3.1 *For any given output level, the LM monopolist always undersupplies product quality as compared to a profit-seeking monopolist operating with the same technology.*

Proof. To demonstrate this claim, it suffices to compare the levels of the marginal cost of quality in the two regimes. This yields the following inequality:

$$x \cdot \frac{\partial P(x,q)}{\partial q} > \frac{x \cdot c(x,q) \cdot \partial P(x,q)/\partial q}{x \cdot P(x,q) - F} \tag{3.10}$$

which is always true, since, dividing by $x \cdot \partial P(x,q)/\partial q$ on both sides, we obtain:

$$1 > \frac{c(x,q)}{x \cdot P(x,q) - F} = \frac{1}{V(x,q)}. \tag{3.11}$$

This completes the proof. ■

The next step consists in evaluating (3.8) against the first order condition obtaining from the planner's problem:

$$\max_q SW(x,q) = \pi(x,q) + CS(x,q), \tag{3.12}$$

that is:

$$\frac{\partial SW(x,q)}{\partial q} = \frac{\partial \pi(x,q)}{\partial q} + \frac{\partial CS(x,q)}{\partial q} = 0 \tag{3.13}$$

or:

$$\frac{\partial SW(x,q)}{\partial q} = -\frac{\partial c(x,q)}{\partial q} + \int_0^x \frac{\partial P(z,q)}{\partial q} dz\,. \tag{3.14}$$

In correspondence of the LM monopoly optimum, the above expression can be written as follows:

$$\frac{\partial SW(x,q)}{\partial q} = -\frac{x \cdot c(x,q) \cdot \partial P(x,q)/\partial q}{x \cdot P(x,q) - F} + \int_0^x \frac{\partial P(z,q)}{\partial q} dz\,. \tag{3.15}$$

The sign of (3.15) is positive iff

$$\frac{1}{x}\int_0^x \frac{\partial P(z,q)}{\partial q} dz > \frac{c(x,q) \cdot \partial P(x,q)/\partial q}{x \cdot P(x,q) - F}\,, \tag{3.16}$$

and conversely. This entails that the LM monopolist may undersupply or oversupply product quality as compared to the social optimum (given the output level), depending upon the level of the willingness to pay for quality characterising the average consumer.

Now notice that

$$\frac{c(x,q) \cdot \partial P(x,q)/\partial q}{x \cdot P(x,q) - F} = \frac{1}{V(x,q)} \cdot \frac{\partial P(x,q)}{\partial q} \tag{3.17}$$

which is surely smaller than $\partial P(x,q)/\partial q$ since $V(x,q)$ must be larger than one. Using (3.17), inequality (3.16) simplifies as follows:

$$\frac{1}{x}\int_0^x \frac{\partial P(z,q)}{\partial q} dz > \frac{1}{V(x,q)} \cdot \frac{\partial P(x,q)}{\partial q} \tag{3.18}$$

revealing that the willingness to pay of the marginal consumer (on the r.h.s.) is weighted by the value added per worker. This feature indeed allows us to state that quality undersupply (as compared to the social optimum) is more likely to obtain under the LM monopoly than under a profit-maximising (PM) monopoly. In the latter case, from Spence (1975) we know that the PM firm distorts quality downwards whenever:

$$\frac{1}{x}\int_0^x \frac{\partial P(z,q)}{\partial q} dz > \frac{\partial P(x,q)}{\partial q} \tag{3.19}$$

i.e., if the willingness to pay for quality of the average consumer is higher than the willingness to pay for quality of the marginal consumer (and conversely). Since

$$\frac{1}{V(x,q)} \cdot \frac{\partial P(x,q)}{\partial q} < \frac{\partial P(x,q)}{\partial q} \tag{3.20}$$

in view of the fact that $V(x,q) > 1$, we may summarise our analysis in the following Proposition:

Proposition 3.2 *Consider a market being alternatively served by a PM firm, an LM firm and a benevolent planner using the same technology. Given the output level,*

(i) if

$$\frac{1}{x}\int_0^x \frac{\partial P(z,q)}{\partial q}dz > \frac{\partial P(x,q)}{\partial q}$$

both the LM and the PM firm undersupply product quality;

(ii) if

$$\frac{1}{x}\int_0^x \frac{\partial P(z,q)}{\partial q}dz \in \left(\frac{\partial P(x,q)}{\partial q}, \frac{1}{V(x,q)} \cdot \frac{\partial P(x,q)}{\partial q}\right)$$

the PM firm oversupplies product quality while the LM firm undersupplies it;

(iii) if

$$\frac{1}{x}\int_0^x \frac{\partial P(z,q)}{\partial q}dz < \frac{1}{V(x,q)} \cdot \frac{\partial P(x,q)}{\partial q}$$

both the PM and the LM firm oversupply product quality.

To illustrate this issue with a familiar example, we can now move on to a more specific layout. First, we summarise the contribution of Lambertini (1997b).

3.4 The linear model

We adopt a simplified version of the vertical differentiation model introduced by Mussa and Rosen (1978). As in the previous section, total costs is $C = wL + F$, where w is the wage rate and L is the amount of labor employed. Here we assume that $L = q^2x$. Normalizing again w to one, we have $C = q^2x + F$. Consumers are uniformly distributed with density 1 over the interval $[\underline{\theta}, \overline{\theta}]$, with $\underline{\theta} > 0$ and $\overline{\theta} = \underline{\theta} + 1$. Consequently, market size is also normalized to one. Parameter θ measures the marginal willingness to pay for quality of the generic consumer, who either buys one unit of the product, or does not buy at all. If he buys, the net surplus is:

$$U = \theta q - p \geq 0, \tag{3.21}$$

otherwise $U = 0$. The corresponding demand function is $x = \overline{\theta} - p/q$. We assume partial market coverage, i.e., $p/q > \underline{\theta}$.

3.4.1 The PM monopolist and the social optimum

From chapter 1, we are already acquainted with the behaviour of the PM monopolist and the social planner, which can be quickly dealt with. The profit-seeking monopolist wants to

$$\max_{p,q} \pi = (p - q^2)x - F. \tag{3.22}$$

Solving this maximum problem yields $p^{PM} = 2\overline{\theta}^2/9$ and $q^{PM} = \overline{\theta}/3$. The associated quantity and profits are $x^{PM} = \overline{\theta}/3$ and $\pi^{PM} = \overline{\theta}^3/27 - F$, respectively. The non-negativity of equilibrium profits requires $F < \overline{\theta}^3/27$. The corresponding level of social welfare is $SW^{PM} = \overline{\theta}^3/18 - F$.

The social planner (SP) must set price and quality so as to maximize social welfare:

$$SW = \int_{p/q}^{\overline{\theta}} (\theta q - q^2) d\theta. \tag{3.23}$$

This objective is attained at $q^{SP} = q^{PM} = \overline{\theta}/3$, i.e., the PM monopolist and the social planner offer the same quality level.[1] However, $p^{SP} = \overline{\theta}^2/9$, with the social planner pricing at marginal cost. Moreover, the socially efficient output is $x^{SP} = 2\overline{\theta}/3 = 2x^{PM}$, and the social welfare level is $SW^{SP} = 2\overline{\theta}^3/27 - F$. In this setting, the PM monopolist exhibits the usual taste for output distortions, while quality is the same as under planning (or perfect competition).[2]

3.4.2 The LM monopolist

The objective of the LM firm consists in maximising:

$$V = \frac{p(\overline{\theta} - p/q) - F}{q^2(\overline{\theta} - p/q)}. \tag{3.24}$$

w.r.t. price and quality. First order conditions are:[3]

$$\frac{\partial V}{\partial p} = \frac{p^2 - qF - 2\overline{\theta}pq + \overline{\theta}^2 q^2}{[q(\overline{\theta}q - p)]^2} = 0; \tag{3.25}$$

[1] This result is driven by the coincidence between the marginal consumer's and the average consumer's valuation for quality (see Spence, 1975, pp. 419-21; and chapter 1).

[2] It is easily shown that the PM monopolist would heavily distort quality if she were compelled to serve the whole market (see chapter 1).

[3] Second order conditions are also met. They are omitted for the sake of brevity.

$$\frac{\partial V}{\partial q} = \frac{2p + pqF - 4\overline{\theta}p^2q - 2F\overline{\theta}q^2 + 2\overline{\theta}^2pq^2}{q^3(\overline{\theta}q - p)^2} = 0\,. \tag{3.26}$$

Solving (3.25-3.26) yields $q^{LM} = 9F/\overline{\theta}^2$ and $p^{LM} = 6F/\overline{\theta}$. The equilibrium output level is $x^{LM} = \overline{\theta}/3$, while the value added per worker amounts to $V^{LM} = \overline{\theta}^3/(27F)$. Social welfare at equilibrium is $SW^{LM} = 3F(1/2 - 9F/\overline{\theta}^3)$. Comparing $\{p^{LM}, q^{LM}, x^{LM}\}$ to $\{p^{PM}, q^{PM}, x^{PM}\}$, one can conclude that the LM monopolist supplies the same quantity as the PM monopolist, although the variety available under LM monopoly is characterized by both a lower quality and a lower price than the variety available under PM monopoly. Both quality and price experience the same reduction, so that the location of the consumer indifferent between buying or not, i.e., the price/quality ratio, is the same as under PM monopoly, and the same consumers are served under both market regimes. The intuition behind the decrease in product quality (as compared to both the social optimum and the PM monopoly) when the LM firm operates in the market is that, since production costs are convex in the quality level while they are linear in the output level, it is in the interest of the LM firm to reduce as much as possible quality rather than quantity, although it remains true that the latter is lower than the socially efficient level. Provided that the costs of quality improvement increase at an increasing rate, any reduction of quality allows for the value added per worker to increase more than proportionally. The two-sided distortion introduced by the LM monopolist considerably reduces social welfare as compared to both the alternative market regimes.

However, the analysis carried out so far in this section refers to global optima, while the discussion contained in the previous section takes output as given. Therefore, we have to step back to the properties of first order conditions w.r.t. quality, for a given output level. To this aim, we write the inverse demand function:

$$p = \left(\overline{\theta} - x\right) q \tag{3.27}$$

according to which the objective functions become:

$$\begin{aligned} V &= \frac{x\left(\overline{\theta} - x\right) q - k}{q^2 x}\,;\ \pi = xq\left(\overline{\theta} - x - q\right) - k \\ SW &= \pi + CS = \frac{xq}{2}\left(2\overline{\theta} - x - 2q\right) - k \end{aligned} \tag{3.28}$$

since consumer surplus is $CS = qx^2/2$. In the three market regimes under consideration, the optimal quality choice is driven by the following conditions, respectively:

$$\begin{aligned}
\frac{\partial V}{\partial q} &= -\frac{x\left(\overline{\theta}-x\right)q-2k}{q^{3}x}=0\Rightarrow q^{V}=\frac{2k}{x\left(\overline{\theta}-x\right)} \\
\frac{\partial \pi}{\partial q} &= x\left(\overline{\theta}-x-2q\right)=0\Rightarrow q^{\pi}=\frac{\overline{\theta}-x}{2} \\
\frac{\partial SW}{\partial q} &= \frac{x}{2}\left(2\overline{\theta}-x-4q\right)=0\Rightarrow q^{SW}=\frac{2\overline{\theta}-x}{4}
\end{aligned} \tag{3.29}$$

where it is immediate to verify that, given x, $q^{SW} > q^{\pi}$, i.e., the PM monopolist undersupplies quality as compared to the social planner, *ceteris paribus*. The comparison between (i) q^{V} and q^{π}; and (ii) q^{V} and q^{SW} is more difficult because of the presence of the fixed cost k in the optimal quality level chosen by the LM monopolist. However, this task can be accomplished by noting that the first derivative of profits w.r.t. q simplifies as follows:

$$\frac{\partial \pi}{\partial q}=\frac{x\left(\overline{\theta}-x\right)^{2}-4k}{\overline{\theta}-x} \tag{3.30}$$

if evaluated at $q = q^{V}$. The above expression is positive for all

$$k<\frac{x\left(\overline{\theta}-x\right)^{2}}{4}. \tag{3.31}$$

Moreover, at $q = q^{V}$, profits are:

$$\pi=\frac{\left[x\left(\overline{\theta}-x\right)^{2}-2k\right]2k}{x\left(\overline{\theta}-x\right)^{2}}-k \tag{3.32}$$

with $\pi > 0$ for all

$$k<\frac{x\left(\overline{\theta}-x\right)^{2}}{4}. \tag{3.33}$$

Hence, the LM monopolist undersupplies product quality as compared to the PM monopolist for any given output level, as long as the above viability condition is met. In view of the previous result concerning the relative performance of the PM monopolist against social planning, we can also conclude that, for any fixed x, the quality offered by the LM firm is clearly socially suboptimal.

3.5 Concluding remarks

We have extended the analysis carried out by Spence (1975) to account for the optimal behaviour of a labour-managed monopolist as far as the supply of product quality is concerned. In general, no clear-cut conclusion can be drawn, as the outcome depends on the distribution of income across the population of consumers. However, in the case of a uniform distribution, there clearly emerges that the maximisation of value added per worker entails that the LM firm distorts quality downwards as compared to both the profit-seeking monopolist and the social planner.

References

1] Champsaur, P. and J.-C. Rochet (1989), "Multiproduct Duopolists", *Econometrica*, **57**, 533-57.
2] Gabszewicz, J.J., A. Shaked, J. Sutton and J.-F. Thisse (1986), "Segmenting the Market: The Monopolist Optimal Product Mix", *Journal of Economic Theory*, **39**, 273-89.
3] Itoh, M. (1983), "Monopoly, Product Differentiation and Economic Welfare", *Journal of Economic Theory*, **31**, 88-104.
4] Lambertini, L. (1997a), "The Multiproduct Monopolist under Vertical Differentiation: An Inductive Approach", *Recherches Economiques de Louvain*, **63**, 109-22.
5] Lambertini, L. (1997b), "On the Provision of Product Quality by a Labor-Managed Monopolist", *Economics Letters*, **55**, 279-83.
6] Maskin, E. and J. Riley (1984), "Monopoly with Incomplete Information", *RAND Journal of Economics*, **15**, 171-96.
7] Mussa, M. and S. Rosen (1978), "Monopoly and Product Quality", *Journal of Economic Theory*, **18**, 301-17.
8] Sheshinski, E. (1976), "Price, Quality and Quantity Regulation in Monopoly Situations", *Economica*, **43**, 127-37.
9] Spence, A.M. (1975), "Monopoly, Quality and Regulation", *Bell Journal of Economics*, **6**, 417-29.

Chapter 4

Endogenous timing in a vertically differentiated duopoly

Luca Lambertini[1]

4.1 Introduction

The way firms can be expected to conduct oligopolistic competition has represented a relevant issue in the economists' research agenda for a long time. The earliest literature in this field treated a relevant feature such as the choice between simultaneous and sequential moves as exogenous (Stackelberg, 1934; Fellner, 1949). Later contributions considered as a sensible approach to investigate the preferences of firms over the distribution of roles in price or quantity games (Gal-Or, 1985; Dowrick, 1986; Boyer and Moreaux, 1987a; 1987b). The preference for leadership (respectively, followership) in quantity (price) games can be established on the basis of the slope of firms' reaction functions or, likewise, noting that products are strategic substitutes (com-

[1]The material contained in this chapter previously circulated under the title "Time Consistency in Games of Timing", Discussion Paper no. 97/10, Institute of Economics, University of Copenhagen. I thank Svend Albæk, Vincenzo Denicolò, Egbert and Hildegard Dierker, Paul Klemperer, Massimo Marinacci and Martin Slater for insightful comments and suggestions. I am also grateful to the audience at the European Meeting of the Econometric Society, Istanbul, August 1996, the XXIII EARIE Conference, Vienna, September 1996 (where a previous version was presented, under the title "Sequential vs Simultaneous Equilibria in a Differentiated Duopoly"), and the Conference "Topics in Microeconomics and Game Theory", Copenhagen, June 1997, for stimulating discussion. The responsibility remains with me only.

plements).[2] Other authors have taken into account the possibility that cost asymmetry or uncertainty may lead to Stackelberg equilibria (Ono, 1982; Albæk, 1990).

The idea that preplay communication can allow agents to play a Stackelberg equilibrium, if there exists at least one dominating the Nash equilibrium (or equilibria) of thc game can be traced back to d'Aspremont and Gérard-Varet (1980). Recent literature explicitly models the strategic choice of timing, which is often possible in reality. Robson (1990a; see also 1989) has proposed an extended duopoly model where price competition takes place in a single period, preceded by firms' scattered price decisions, which cannot be altered. Only Stackelberg equilibria emerge from such a game. In an influential paper, Hamilton and Slutsky (1990) investigated the endogenous choice of roles, i.e., the endogenous arising of Stackelberg or Cournot equilibria, in noncooperative two-person games (typically, duopoly games), by analysing an extended game where players (say, firms) are required to set both the actual moves or actions and the time at which such actions are to be implemented. Their approach is close in spirit to Robson's (1990a), though they also consider Cournot competition and the mixed case where one firm sets her price and the other firm decides her output level. Indeed, Hamilton and Slutsky's model shares many features with Robson (1989). When firms choose to act at different times, sequential equilibria obtain, while if they decide to move at the same time, simultaneous Nash equilibria are observed. The choice of the timing occurs in a preplay stage which does not take place in real time, so that there is no discounting associated with waiting and payoffs are the same whether firms choose to move as soon as possible or delay as long as they can. The decision to play early or at a later time is not sufficient *per se* to yield sequential play, since an analogous decision taken by the rival leads to simultaneous play.

Hamilton and Slutsky (1990, HS henceforth) assume that each of the games associated with simultaneous or sequential play has a unique equilibrium. The immediate consequence of this hypothesis is a lemma according to which each firm strictly prefers her payoff as a leader to that accruing to her under simultaneous moves. Building on such a lemma, HS show that a Stackelberg equilibrium with sequential play is selected as a subgame perfect equilibrium of the extended game with observable delay if and only if the outcome of sequential play Pareto-dominates the outcome associated with simultaneous play (HS, 1990, Theorems III and IV). Otherwise, if firms are better off playing simultaneously rather than accepting the follower's role,

[2] The concept of strategic substitutability/complementarity is due to Bulow, Geanakoplos and Klemperer (1985).

the subgame perfect equilibrium involves simultaneous play (HS, 1990, Theorem II). Summing up, in a game where firms choose a single variable, their respective reaction functions are monotone in the rival's strategic variable and unique and distinct Nash and Stackelberg equilibria exist in the interior of the action space, (a) if both reaction functions have the same slope, then alternatively (i) neither intersects the Pareto-superior set, in which case the timing game has a unique equilibrium involving simultaneous play, or (ii) both reaction functions intersect the Pareto-superior set, in which case both Stackelberg equilibria are equilibria of the timing game; (b) if reaction functions have opposite slopes, the timing game has a unique equilibrium where the firm whose reaction function intersects the Pareto-superior set moves second (HS, 1990, Theorem V).[3] Recently, Amir (1995) has provided a counterexample to HS's Theorem V, showing that the monotonicity of best-reply functions is insufficient for HS's Theorem V to hold, and the characterization of the order of moves in the extended game requires the monotonicity of each player's (or firm's) payoff function in the rival's actions.

The possible consequences of asymmetric information on the order of moves are accounted for by Mailath (1993). In a quantity game with asymmetric information about demand, he shows that the informed firm does not exploit her chance to move before the rival. Pal (1996) explicitly takes into account mixed strategies. He considers an extended quantity-setting game with two identical firms and two production periods before the market-clearing instant. He shows that only three outcomes are possible: (i) both firms produce in the second period, so that a simultaneous Cournot equilibrium obtains; (ii) firms produce in different periods, yielding a Stackelberg-like equilibrium (see also Robson, 1990b); (iii) Stackelberg warfare may arise when firms produce in the first period, but both produce more than in the Cournot-Nash equilibrium.

The aim of this paper is threefold. First, I shall extend the analysis provided by HS by showing that their box of tools can be profitably used in a more general environment than the one they have described. Specifically, I am going to prove that sequential play obtains at the subgame perfect equilibrium of an extended game with observable delay if and only if both the leader and the follower are at least weakly better off than under simultaneous play.

[3] HS (1990, section IV) also consider an extended game with action commitment in the spirit of Dowrick (1986), where each firm must commit to a particular action irrespective of the rival trying to lead or follow. This yields multiple equilibria with both simultaneous and sequential play. The extension of this approach to multi-period m-player settings is in Matsumura (1999), who shows that, at the subgame perfect equilibrium, at least $m-1$ players take the lead and at most one follows.

Second, I will analyse a two-stage game of vertical differentiation where firms choose the timing of moves, product quality and compete *à la* Bertrand on the market, allowing for the payoff sequence to be such that the leader's payoff is not necessarily preferred to the simultaneous play payoff. Two different extended games can be conceived. In the first, firms take their timing decisions between the quality and the price stage. Here, on the basis of strategic complementarity between prices, as well as the normal form of the game, it emerges that firms decide to play sequentially. In the alternative extended game, the timing decisions are taken before playing both stages taking place in real time. Here, provided firms can irreversibly commit to their respective timing decisions, unusual results may emerge in terms of preferences over the distribution of roles. The subgame perfect equilibrium of such an extended game drastically differs from that observed under price competition when firms cannot endogenously differentiate their respective goods. Specifically, in the game I present, simultaneous play emerges when firms bear variable production costs, due to the fact that the price leader's profit is lower than simultaneous play profit, so that both duopolists play at the latest opportunity in order to avoid being first. Otherwise, when costs take the form of R&D efforts, a sequential equilibrium emerges with the low-quality firm taking the lead. This entails that most oligopoly models where market competition is preceded by a stage in which firms proceed to take a commitment that affects the ensuing price or quantity subgame (quality choice, location, delegation, or R&D) are likely to produce equilibrium outcomes where preferences over the distribution of roles are drastically different as compared to one-stage games where price or quantity is the only strategic variable.

Third, I investigate the time (in)consistency of the timing choices, which is directly related to the possibility for players to locate the timing stage at any point along the game tree, whenever the game unravels in more than one stage. The issue of time (in)consistency is well known since Kydland and Prescott (1977). The original formulation of the problem refers to a policy game between government and economic agents endowed with rational expectations. Time inconsistency materialises into the incentive for the policy-maker to renege *ex post* a given policy announcement. In general, one can think that in any given multi-stage game there may exist a problem of time inconsistency when at least one of the players has an incentive to renege *ex post* the announcement pertaining to an action to be undertaken at a later stage. In this paper, I prove that a sufficient condition for the extended game with observable delay to be time consistent is that the timing stage immediately precede the stage which the choice of timing refers to.

Using the vertical differentiation model as an example allows several other conclusions, although limited to a specific setting. The two different extended games with observable delay are characterized by different subgame perfect equilibria, determined by different sequences of moves. Hence, changing the location of the timing stage drastically changes the outcome of the extended game. It appears that, to be consistent (and thus also credible), timing announcements made before any move in real time need to be supported by a commitment technology forcing firms to stick to such announcements once they reach the price stage. Otherwise, if such devices are not available, at the price stage any timing combination that does not yield sequential play is not credible. Therefore, to avoid time inconsistency the extension concerning timing decisions must be located between the first and the second stage of the basic game. It emerges that the choice of timing in multistage games can jeopardize HS's conclusions, in a way that recalls the point raised by Amir (1995). This has a last straightforward implication for multistage games. If players are required to set the timing of their respective moves at a particular stage, then locating the timing decision just upstream of that stage will always avoid problems of time inconsistency.

The remainder of the paper is structured as follows. The generalization of the extended game approach is discussed in section 4.2. Section 4.3 is devoted to the description of the vertical differentiation setting. Sections 4.4 and 4.5 describe the extended games that can be envisaged under vertical differentiation. The issue of time consistency is then dealt with in section 4.6. Finally, section 4.7 contains concluding remarks.

4.2 Extended games with observable delay

Consider the extension of a two-stage game where firms can set a strategic variable (price or quantity) in the downstream stage and another variable (e.g., the R&D effort, product quality, location, etc.) in the upstream stage.[4] Then, as in HS, the extension consists in choosing noncooperatively between moving first or second in the downstream market stage only, while moves are simultaneous in the upstream stage. I shall adopt here a symbology which largely replicates that in HS (1990, p. 32). Two different extended games are considered. In the first, the timing decisions pertaining to the moves in the second stage of the basic game are taken between the first and the second stage of the basic game. In the second extended game, the timing decisions are taken before any decision in real time takes place, that is, before deciding

[4] Both d'Aspremont and Gérard-Varet (1980) and Hamilton and Slutsky (1990) consider one-stage games. In examining multistage games, I assume perfect recall. For the analysis of time consistency in games with imperfect recall, see Battigalli (1997). See also Piccione and Rubinstein (1997a,b).

upon the variables pertaining to both stages forming the basic game.

4.2.1 The first extended game with observable delay

Define as $\overline{\Gamma} = (N, \Sigma, \overline{\Pi})$ the first extended game with observable delay, where the extension takes place between the first and the second stage of the basic game. The set of players (or firms) is $N = \{A, B\}$, and $\alpha(\xi_A, \xi_B)$ and $\beta(\xi_A, \xi_B)$ are the compact and convex intervals of $\mathbb{R}$ representing the actions available to agents A and B in the downstream stage, conditional upon the choices made in the upstream stage where they are required to set ξ_A and ξ_B, respectively. $\overline{\Pi}$ is the payoff function, such that individual payoffs are defined as $\pi_A(\overline{\xi}_A, \overline{\xi}_B) : \alpha(\overline{\xi}_A, \overline{\xi}_B) \times \beta(\overline{\xi}_A, \overline{\xi}_B) \rightarrow \mathbb{R}$ and $\pi_B(\overline{\xi}_A, \overline{\xi}_B) : \alpha(\overline{\xi}_A, \overline{\xi}_B) \times \beta(\overline{\xi}_A, \overline{\xi}_B) \rightarrow \mathbb{R}$. The bar indicates that $\overline{\xi}_A$ and $\overline{\xi}_B$ are a generic given pair which may or may not (and as a general rule they do not) coincide with the subgame perfect values of ξ_A and ξ_B, as determined by backward induction when one takes into account the timing chosen in the downstream stage. I assume that, for any given pair $(\overline{\xi}_A, \overline{\xi}_B)$, π_i is single-valued in the action chosen by player j. The set of times at which firms can choose to move is $T = \{F, S\}$, i.e., *first* or *second*. The set of strategies for player i is $\Sigma_i = \{F, S\} \times \Phi_i$, where Φ_i maps $T \times \beta(\overline{\xi}_A, \overline{\xi}_B)$ (or $\alpha(\overline{\xi}_A, \overline{\xi}_B)$) into $\alpha(\overline{\xi}_A, \overline{\xi}_B)$ (or $\beta(\overline{\xi}_A, \overline{\xi}_B)$).

If in the market subgame both firms choose to move at the same time (F-F or S-S), they obtain the payoffs associated with the simultaneous Nash equilibrium, $(\pi^n_A(\overline{\xi}_A, \overline{\xi}_B), \pi^n_B(\overline{\xi}_A, \overline{\xi}_B))$, otherwise they get the payoffs associated with the Stackelberg equilibrium, e.g., $(\pi^l_A(\overline{\xi}_A, \overline{\xi}_B), \pi^f_B(\overline{\xi}_A, \overline{\xi}_B))$ if A moves first and B moves second, or *vice versa*. Superscript n denotes the Nash equilibrium, while superscripts l and f denote *leader* and *follower*, respectively. Define the set of pure-strategy equilibria at the timing stage as $\overline{\Omega} = \{(T_A(\overline{\xi}_A, \overline{\xi}_B), T_B(\overline{\xi}_A, \overline{\xi}_B))\}$.

4.2.2 The second extended game with observable delay

Define as $\Gamma^* = (N, \Sigma, \Pi^*)$ the second extended game with observable delay. Again, the set of players (or firms) is $N = \{A, B\}$, and $\alpha(\xi_A, \xi_B)$ and $\beta(\xi_A, \xi_B)$ are the compact and convex intervals of $\mathbb{R}$ representing the actions available to agents A and B in the downstream stage, conditional upon the choices made in the upstream stage where they are required to set ξ_A and ξ_B, respectively. Π^* is now the payoff function, such that individual payoffs are defined as $\pi_A(\xi^*_A, \xi^*_B) : \alpha(\xi^*_A, \xi^*_B) \times \beta(\xi^*_A, \xi^*_B) \rightarrow \mathbb{R}$ and $\pi_B(\xi^*_A, \xi^*_B) : \alpha(\xi^*_A, \xi^*_B) \times \beta(\xi^*_A, \xi^*_B) \rightarrow \mathbb{R}$. The star indicates that the choice of ξ_A and ξ_B (which firms accomplish through simultaneous moves) is part of the subgame perfect equilibrium path which is determined by backward induction when one takes into account the timing chosen in the downstream stage. The set of times at which firms can choose to move is $T = \{F, S\}$, i.e., *first* or *second*. The set of strategies for player i is $\Sigma_i = \{F, S\} \times \Phi_i^*$, where Φ_i^* maps $T \times \beta(\xi_A^*, \xi_B^*)$ (or $\alpha(\xi_A^*, \xi_B^*)$) into $\alpha(\xi_A^*, \xi_B^*)$ (or $\beta(\xi_A^*, \xi_B^*)$).

If in the market subgame both firms choose to move at the same time (F-F or S-S), they obtain the payoffs associated with the simultaneous Nash equilibrium, $(\pi_A^n(\xi_A^n, \xi_B^n), \pi_B^n(\xi_A^n, \xi_B^n))$, otherwise they get the payoffs associated with the Stackelberg equilibrium, e.g., $(\pi_A^l(\xi_A^l, \xi_B^f), \pi_B^f(\xi_A^l, \xi_B^f))$ if A moves first and B moves second, or *vice versa*. The superscripts n, l, and f associated with ξ_A and ξ_B indicate that the values of these variables are chosen optimally, according to the shape of downstream competition. Finally, define the set of pure-strategy equilibria at the timing stage as $\Omega^* = \{(T_A(\xi_A^*, \xi_B^*), T_B(\xi_A^*, \xi_B^*))\}$.

Both games can be described in normal form as in matrix 4.1, where $(\cdot, \cdot)$ stands either for $(\overline{\xi}_A, \overline{\xi}_B)$ or for the relevant (ξ_A^*, ξ_B^*).

		B	
		F	S
A	F	$\pi_A^n(\cdot,\cdot), \pi_B^n(\cdot,\cdot)$	$\pi_A^l(\cdot,\cdot), \pi_B^f(\cdot,\cdot)$
	S	$\pi_A^f(\cdot,\cdot), \pi_B^l(\cdot,\cdot)$	$\pi_A^n(\cdot,\cdot), \pi_B^n(\cdot,\cdot)$

Matrix 4.1: The reduced-form extended game

Notice that, in the absence of the upstream stage where firms must set ξ_A and ξ_B, this game coincides with that considered by HS, so that matrix 4.1 would collapse into their matrix (cf. HS, 1990, p. 33). In the remainder of the paper, I will assume what follows:

Assumption 4.1 *Both $\overline{\Omega}$ and Ω^* are non-empty.*

Assumption 4.1 rules out situations like the one that would arise if payoffs in matrix 1 were ranked as follows: $\pi_A^n(\cdot,\cdot) > \pi_A^l(\cdot,\cdot) > \pi_A^f(\cdot,\cdot)$; $\pi_B^l(\cdot,\cdot) > \pi_B^f(\cdot,\cdot) > \pi_B^n(\cdot,\cdot)$.

HS (1990, p. 31) assume that each of the basic games generated by a particular timing combination has a unique equilibrium, and that these differ from each other. Then, on this basis, HS (1990, Lemma I, p. 35) show that each player's (firm's) leadership payoff must exceed his payoff in simultaneous play because if he is the leader, he is obviously able to choose the best position along the follower's reaction function, so that the Nash equilibrium point is feasible for him. If he accepts to move first (and chooses a point which differs from the Nash equilibrium one), it must be true that he is at least as well off as in the simultaneous equilibrium. *Per se*, this argument appears intuitive and unquestionable. Though, intuition also suggests that analogous considerations must hold for the follower as well. Consider a firm that is contemplating the opportunity of moving second. Provided that by moving at the first occasion, she can at least obtain the Nash payoff, she will accept moving late only if she is better off as a follower than in any other situation. Notice that this is precisely what emerges from HS's Theorems II and III. Accordingly, I state the following:

Lemma 4.1 *A necessary condition for sequential play in pure strategies to emerge at the subgame perfect equilibrium of the extended game with observable delay is that each player's leadership payoff be higher than his payoff under simultaneous play.*

and

Lemma 4.2 *A necessary condition for sequential play in pure strategies to emerge at the subgame perfect equilibrium of the extended game with observable delay is that there exists at least one Stackelberg equilibrium where the follower's payoff be higher than his payoff under simultaneous play.*

Considered jointly, lemma 4.1 and lemma 4.2 yield a *necessary and sufficient* condition for sequential play to obtain at the subgame perfect equilibrium of the extended game with observable delay. This is stated in

Proposition 4.1 *The subgame perfect equilibrium of the extended game with observable delay involves sequential moves if and only if the basic game exhibits at least one Stackelberg equilibrium that Pareto-dominates the simultaneous Nash equilibrium.*

In other words, the method for equilibrium selection proposed by HS holds with no specific requirement on the sequence of profits associated with the roles firms can play in the basic game. Note that this picture largely replicates the notion of Stackelberg-solvable game as defined by d'Aspremont and Gérard-Varet (1980, Theorem 1.1, p. 203). The same holds w.r.t. Theorem 1 in Robson (1989, pp. 919-20).

As to the issue of time (in)consistency, I introduce the following definitions:

Definition 4.1 *An extended game with observable delay is strictly time consistent if $\Omega^* \equiv \overline{\Omega}$.*

Definition 4.2 *An extended game with observable delay is weakly time consistent if $\Omega^* \subset \overline{\Omega}$.*

Definition 4.3 *An extended game with observable delay is weakly time inconsistent if $\Omega^* \cap \overline{\Omega} \neq \emptyset$.*

Definition 4.4 *An extended game with observable delay is time inconsistent if $\Omega^* \cap \overline{\Omega} = \emptyset$.*

This amounts to saying that an extended game is

I. strictly time consistent, if the location of the timing choice along the game tree is irrelevant as to the set of pure-strategy equilibria;

II. weakly time consistent, if the set of equilibria of the game where the timing choice takes place before any other stage is played in real time, is a proper subset of the set of equilibria observed if the timing choice is located just upstream the stage to which it refers;

III. weakly time inconsistent, if the intersection between the two sets is non-empty and Ω^* is not a subset of $\overline{\Omega}$;

IV. time inconsistent, if the two equilibria sets' intersection is empty.

A first result is stated in the following:

Proposition 4.2 *A sufficient condition for an extended game with observable delay to be strictly time consistent is that the timing choice immediately precede the stage which the choice itself refers to.*

The proof consists in the following intuitive argument. Suppose that the timing choice is contiguous to the stage which it refers to. If so, we have indeed that $\xi_K^* = \overline{\xi}_K$, for $K = A, B$. Then, it follows necessarily that $\Omega^* \subseteq \overline{\Omega}$. However, it is fairly obvious we may obtain $\Omega^* \subseteq \overline{\Omega}$ without imposing the aforementioned condition.

This raises another question. If players' timing decisions are unaffected by the location of the timing stage itself, the extended game is time consistent and timing announcements are indeed credible. Moreover, the two versions of the game are observationally equivalent and firms can disregard the issue of which kind of game they are actually playing. If, instead, it is in the interest of at least one player to renege *ex post* a previous declaration in at least one of the two extended games proposed here, then the two games are not observationally equivalent. This entails that announcements by at least one firm are not credible,[5] and a further issue arises, namely, which one of the two games will be endogenously selected by firms.

These issues are investigated in the following sections, where I adopt a two-stage model describing a duopoly market for vertically differentiated goods.

4.3 A differentiated duopoly model

Consider a duopolistic market where firms supply a vertically differentiated good, whose quality is denoted by q_i, $i = H, L$, with $q_H \geq q_L > 0$. They employ the same productive technology, which can alternatively involve variable costs of quality improvements,

$$C_i = q_i^2 x_i, \tag{4.1}$$

where x_i denotes the output of firm i; or fixed costs of quality improvements,

$$C_i = q_i^2, \tag{4.2}$$

which may be the case when the cost of increasing the quality level falls on R&D investments and is not related to the scale of production (see, *inter alia*, Gabszewicz and Thisse, 1979; Shaked and Sutton, 1982, 1983; Lehmann-Grube, 1997; Lambertini, 1999).

Consumers are uniformly distributed over the interval $[0, \overline{\theta}]$. Parameter θ represents each consumer's marginal willingness to pay for quality, and it can be thought of as the reciprocal of the marginal utility of nominal income or money (cf. Tirole, 1988, p. 96). As $\overline{\theta}$ increases, the size of the market increases. Consumers' density can be normalised to one, so that total population is also equal to one. The indirect utility function of the generic consumer is:

$$U = \theta q_i - p_i. \tag{4.3}$$

If the consumer buys, he buys just one unit of the product from the firm that offers the price-quality ratio ensuring the highest utility. Let h and k denote

[5] In this situation, players (e.g., firms) would be able to stick to previous announcements if and only if a commitment technology were available, as, e.g., a given productive capacity (see Kreps and Scheinkman, 1983; Davidson and Deneckere, 1986, *inter alia*). Observe that, under this respect, the issue of time (in)consistency as defined here is the same as in the seminal paper by Kydland and Prescott (1977).

the marginal willingness to pay characterizing, respectively, the consumer who is indifferent between the high and the low-quality good, and that who is indifferent between buying the low-quality good or nothing at all:

$$h = \frac{p_H - p_L}{q_H - q_L}; \quad k = \frac{p_L}{q_L}. \tag{4.4}$$

Then, the market demands for the two varieties are, respectively,

$$x_H = \overline{\theta} - .h \text{ if } h \in]k, \overline{\theta}[, \tag{4.5}$$

$$x_L = h - k \text{ if } k \in]0, h[. \tag{4.6}$$

By inverting the system (4.5-4.6), one obtains the demand functions pertaining to Cournot behavior:

$$p_H = \overline{\theta} q_H - q_H x_H - q_L x_L, \tag{4.7}$$

$$p_L = q_L(\overline{\theta} - x_H - x_L). \tag{4.8}$$

Competition takes place in two stages, the first played in the quality space, the second either in the price or in the quantity space. In the first extended game, the extension takes place between the quality stage and the market stage, so that the relevant payoff facing firms when they are required to set the timing of moves pertaining to the market stage are the profit functions defined for a generic pair of qualities $(\overline{q}_H, \overline{q}_L)$. In the second extended game the extension precedes both stages and the relevant payoffs are the equilibrium profits obtained in correspondence of the specific pair (q_H^*, q_L^*) which is optimal given the sequence of moves in the market stage.

4.4 The first extended game with observable delay

In this section I consider the case where the extension concerning the choice of timing is inserted between the quality stage and the market stage, so that in choosing whether to move early or late firms face a matrix where profits are defined as a function of a generic pair of quality levels. The main aim of the following analysis is to establish that π_H (resp., π_L) is single-valued in L's (resp., H) price choice for any given quality pair, in the admissible range of $\overline{\theta}$.

4.4.1 Variable costs of quality improvement: Bertrand competition

Assume production costs are described by (1). Firms' objective functions are defined as follows:

$$\pi_H = (p_H - q_H^2)x_H;\ \pi_L = (p_L - q_L^2)x_L. \tag{4.9}$$

A preliminary observation concerning the viable quality range is that, given (4.1) and (4.3), any change in the quality level produced by either firm must respect the condition $\theta dq_i \geq 2q_i dq_i$. Since the upper bound of θ is $\overline{\theta}$, the latter inequality implies $q_i \in]0, \overline{\theta}/2]$ (cf. Delbono, Denicolò and Scarpa, 1996, p. 36). This information will be useful below. The game is solved by backward induction. Consider first the fully simultaneous game. The first order conditions (FOCs) at the price stage are:

$$\frac{\partial \pi_H}{\partial p_H} = \frac{p_L - 2p_H + \overline{\theta} q_H - \overline{\theta} q_L + q_H^2}{q_H - q_L} = 0; \tag{4.10}$$

$$\frac{\partial \pi_L}{\partial p_L} = \frac{p_H q_L - 2p_L q_H + q_H q_L^2}{q_L(q_H - q_L)} = 0. \tag{4.11}$$

The above FOCs implicitly define increasing reaction functions in the price space, i.e., as it is usually observed under price competition, there exists strategic complementarity (Bulow, Geanakoplos and Klemperer, 1985). Solving the system (4.10-4.11), I obtain the equilibrium prices:

$$p_H^n = \frac{q_H}{4q_H - q_L}[2\overline{\theta}(q_H - q_L) + 2q_H^2 + q_L^2]; \tag{4.12}$$

and

$$p_L^n = \frac{q_L}{4q_H - q_L}[\overline{\theta}(q_H - q_L) + q_H(q_H + 2q_L)], \tag{4.13}$$

where superscript n stands for Nash equilibrium. This yields the following Bertrand-Nash equilibrium profits:

$$\pi_H^n = \frac{q_H^2(q_H - q_L)(2\overline{\theta} - 2q_H - q_L)^2}{(4q_H - q_L)^2};\ \pi_L^n = \frac{q_H q_L(q_H - q_L)(\overline{\theta} + q_H - q_L)^2}{(4q_H - q_L)^2}. \tag{4.14}$$

The leader's problem in the price stage can be described as follows:

$$\max_{p_i} \pi_i \tag{4.15}$$

$$s.t.: \frac{\partial \pi_j}{\partial p_j} = 0, i \neq j, \tag{4.16}$$

for both firms, i.e., it consists in the maximization of the leader's profit under the constraint represented by the follower's reaction function, implicitly given by the derivative of her profit function w.r.t. her price.[6] The equilibrium prices that obtain in the two problems can be found in Appendix 4.1. Equilibrium profits under sequential play are

$$\pi^l_H = \frac{q_H(q_H - q_L)(2\overline{\theta} - 2q_H - q_L)^2}{8(2q_H - q_L)^2}; \quad \pi^f_L = \frac{q_H q_L(q_H - q_L)(2\overline{\theta} + q_H - 3q_L)^2}{16(2q_H - q_L)^2}; \tag{4.17}$$

$$\pi^f_H = \frac{(q_H - q_L)(4\overline{\theta}q_H - \overline{\theta}q_L - 4q_H^2 - q_H q_L + q_L^2)^2}{16(2q_H - q_L)^2}; \quad \pi^f_L = \frac{q_L(q_H - q_L)(\overline{\theta} + q_H - q_L)^2}{8(2q_H - q_L)^2}. \tag{4.18}$$

It can be easily established that $\pi^f_H \geq \pi^l_H \geq \pi^n_H$ and $\pi^l_L \geq \pi^n_L$ for all $q_H \geq q_L > 0$. As to the comparison between the leadership profit and the followership profit for the low-quality firm, one obtains the following

$$\text{sign } (\pi^f_L - \pi^l_L) = \text{ sign } (2\overline{\theta}^2 - 4\overline{\theta}q_L - 2q_H^2 + q_H q_L + 2_L^2). \tag{4.19}$$

The roots w.r.t. q_H of the polynomial in (4.19) are

$$q_{H1} = \frac{q_L - \sqrt{16\overline{\theta}^2 - 32\overline{\theta}q_L + 17q_L^2}}{4}; \quad q_{H2} = \frac{q_L + \sqrt{16\overline{\theta}^2 - 32\overline{\theta}q_L + 17q_L^2}}{4}, \tag{4.20}$$

where $q_{H1} \leq 0$ and $q_{H2} \in [0.64039\overline{\theta}, \overline{\theta}]$ for $q_L \in [0, \overline{\theta}/2]$. Hence, $\pi^f_L \geq \pi^l_L$ for all $q_H \in [q_L, \overline{\theta}/2]$. This entails that $\pi^f_L \geq \pi^l_L \geq \pi^n_L$ for all admissible quality levels. Consequently, the set of pure-strategy equilibria is $\overline{\Omega} = \{(F_H(\overline{q}_H, \overline{q}_L), S_L(\overline{q}_H, \overline{q}_L)), (F_H(\overline{q}_H, \overline{q}_L), S_L(\overline{q}_H, \overline{q}_L))\}$.

[6] Several other equilibria could be investigated, if firms were assumed to be able to play sequentially also in the quality stage, or set quantities instead of prices in the market stage. For an analysis of such equilibria, see Lambertini (1996).

4.4.2 Variable costs of quality improvement: Cournot competition

Again, assume production costs are given by (4.1).[7] When firms set output levels simultaneously, the Cournot-Nash equilibrium at the market stage is the solution of the following FOCs:

$$\frac{\partial \pi_H}{\partial x_H} = \overline{\theta} q_H - q_H^2 - 2q_H x_H - q_L x_L = 0; \tag{4.21}$$

$$\frac{\partial \pi_L}{\partial x_L} = q_L(\overline{\theta} - x_H - x_L) - q_L x_L - q_L^2 = 0, \tag{4.22}$$

yielding

$$x_H^n = \frac{2\overline{\theta} q_H - 2q_H^2 - \overline{\theta} q_L + q_L^2}{4q_H - q_L}; \quad x_L^n = \frac{q_H(\overline{\theta} + q_H - 2q_L)}{4q_H - q_L}. \tag{4.23}$$

Plugging (4.23) into firms' objective functions, I obtain the Cournot-Nash equilibrium profits defined in terms of a generic quality pair:

$$\pi_H^n = \frac{q_H(2\overline{\theta} q_H - 2q_H^2 - \overline{\theta} q_L + q_L^2)^2}{(4q_H - q_L)^2}; \quad \pi_L^n = \frac{q_H^2 q_L(\overline{\theta} + q_H - 2q_L)^2}{(4q_H - q_L)^2}. \tag{4.24}$$

When sequential play is adopted, the leader's problem is as in (4.15-4.16), yielding

$$\begin{aligned} \pi_H^l &= \frac{(2\overline{\theta} q_H - 2q_H^2 - \overline{\theta} q_L + q_L^2)^2}{2(2q_H - q_L)}; \\ \pi_L^f &= \frac{q_L(2\overline{\theta} q_H + 2q_H^2 - \overline{\theta} q_L - 4q_H q_L + q_L^2)^2}{16(2q_H - q_L)^2}; \end{aligned} \tag{4.25}$$

$$\begin{aligned} \pi_H^f &= \frac{q_H(4\overline{\theta} q_H - 4q_H^2 - 3\overline{\theta} q_L + q_H q_L + 2q_L^2)^2}{16(2q_H - q_L)^2}; \\ \pi_L^l &= \frac{q_H q_L(\overline{\theta} + q_H - 2q_L)^2}{8(2q_H - q_L)}. \end{aligned} \tag{4.26}$$

The output levels corresponding to the two Stackelberg equilibria can be found in Appendix 4.2. It can be quickly checked that $\pi_i^l > \pi_i^n > \pi_i^f$, $i = H, L$, for all $\overline{\theta} \geq q_H \geq q_L > 0$. Given that the viable range for q_i is $\left(0, \overline{\theta}/2\right]$, the above profit ranking holds everywhere. As a result, the set of pure-strategy equilibria is $\overline{\Omega} = \{(F_H(\overline{q}_H, \overline{q}_L), F_L(\overline{q}_H, \overline{q}_L))\}$. According to

[7] The game where Cournot competition follows a product stage where quality improvements are obtained through a fixed cost is not described in that it yields the same results in terms of the choice of timing.

the definition of d'Aspremont and Gérard-Varet (1980, pp. 204-207), the quantity game is *strictly competitive*, meaning that since both firms aim at being the leader, the game cannot be played simultaneously even with preplay communication, or, as is the case here, with a preplay stage where timing is noncooperatively decided upon.

4.4.3 Fixed costs of quality improvement: Bertrand competition

Production costs are given by (4.2), and can be thought of as R&D efforts. Market demands correspond to (4.5-4.6). Firms' profit functions can be written as:

$$\pi_i = p_i x_i - q_i^2, i = H, L. \tag{4.27}$$

Consider first the fully simultaneous game. Proceeding backwards, I calculate the FOCs pertaining to the price stage:

$$\frac{\partial \pi_H}{\partial p_H} = \overline{\theta} - \frac{p_H - p_L}{q_H - q_L} = 0; \tag{4.28}$$

$$\frac{\partial \pi_L}{\partial p_L} = \frac{p_H - p_L}{q_H - q_L} - 2\frac{p_L}{q_L} = 0. \tag{4.29}$$

Again, the FOCs reveal strategic complementarity between prices. Solving the system (4.28-4.29) yields the following equilibrium prices:

$$p_H^n = 2\overline{\theta} q_H \frac{(q_H - q_L)}{(4q_H - q_L)}; \; p_L^n = \overline{\theta} q_L \frac{(q_H - q_L)}{(4q_H - q_L)}. \tag{4.30}$$

The corresponding profits at the quality stage are:

$$\pi_H^n = \frac{4(\overline{\theta} q_H)^2 (q_H - q_L)}{(4q_H - q_L)^2} - q_H^2; \; \pi_L^n = \frac{\overline{\theta}^2 q_H q_L (q_H - q_L)}{(4q_H - q_L)^2} - q_H^2. \tag{4.31}$$

As in the previous subsections, in the cases where sequential play is adopted, the leader's problem is described by (4.15-4.16), yielding

$$\pi_H^l = \frac{\overline{\theta}^2 q_H (q_H - q_L)}{2(2q_H - q_L)} - q_H^2; \; \pi_L^f = \frac{\overline{\theta}^2 q_H q_L (q_H - q_L)}{4(2q_H - q_L)^2} - q_L^2; \tag{4.32}$$

$$\pi_H^f = \frac{\overline{\theta}^2 (4q_H - q_L)^2 (q_H - q_L)}{16(2q_H - q_L)^2} - q_H^2; \; \pi_L^l = \frac{\overline{\theta}^2 q_L (q_H - q_L)}{8(2q_H - q_L)} - q_L^2, \tag{4.33}$$

with $\pi_i^f \geq \pi_i^l \geq \pi_i^n$ for all admissible quality levels. Equilibrium prices can be found in Appendix 4.3. Again, the set of pure-strategy equilibria is $\overline{\Omega} = \{(F_H(\overline{q}_H, \overline{q}_L), S_L(\overline{q}_H, \overline{q}_L)), (F_H(\overline{q}_H, \overline{q}_L), S_L(\overline{q}_H, \overline{q}_L))\}$.

As a consequence, regardless of the technology, in both games both Stackelberg outcomes dominate the simultaneous play outcome, so that the whole discussion above can be summarized in the following:

Proposition 4.3 *The subgame perfect equilibrium of the extended game where the choice of timing is taken between the quality and the price stage involves sequential play, independently of the technology adopted by firms. If instead the market stage is played in the quantity space, the extended game has a unique equilibrium involving simultaneous play.*

This obviously implies that each Bertrand game has also a correlated equilibrium and, finally, a mixed-strategy equilibrium where firms randomize over playing early or delay, hence with a strictly positive probability of moving simultaneously.

4.5 The second extended game with observable delay: irreversible commitment

I am now in a position to consider the game where the timing decision concerning the market stage takes place before the choice of qualities. Assume for the moment that firms can irreversibly commit to the timing decision. To begin with, I describe the setting where production technology involves variable costs.

4.5.1 Variable costs of quality improvement: Bertrand competition

From the FOCs (4.10-4.11) we can observe that, by choosing quality levels in the upstream stage of the game, the high-quality firm affects the intercept while the low-quality firm affects the slope of their respective reaction functions in the price space. This phenomenon is responsible for the outcomes I will illustrate below. The solution of the first stage of the game involves numerical calculations. Normalising $\overline{\theta}$ to one, it can be shown that $q_H^n = 0.40976$ and $q_L^n = 0.19936$.[8] The corresponding equilibrium profits

[8] Here, as well as in the remainder of the section, one should prove that neither firm has any incentive to leapfrog the rival. One such proof has been provided by Motta (1993)

amount to $\pi_H^n = 0.0164$ and $\pi_L^n = 0.0121$. Further numerical computations show that

$$q_i^n(\overline{\theta}) = \overline{\theta} q_i^n(1);\ \pi_i^n(\overline{\theta}) = \overline{\theta}\pi_i^n(1). \tag{4.34}$$

This holds independently of the timing of moves firms adopt in the market stage.

Obviously, equilibrium qualities are different in each of the games being considered, though the quality stage is always played simultaneously. The optimal qualities selected when the high-quality firm is appointed the leader's role in the ensuing price stage are:

$$q_H^l(\overline{\theta}) = \overline{\theta} q_H^l(1) = 0.41601\overline{\theta};\quad q_L^f(\overline{\theta}) = \overline{\theta} q_L^f(1) = 0.21887\overline{\theta}, \tag{4.35}$$

and the corresponding equilibrium profits amount to:

$$\pi_H^l(\overline{\theta}) = \overline{\theta}^3 \pi_H^l(1) = 0.01506\overline{\theta}^3;\quad \pi_L^f(\overline{\theta}) = \overline{\theta}^3 \pi_L^f(1) = 0.01412\overline{\theta}^3. \tag{4.36}$$

It is immediate to verify that (i) both qualities increase as compared to the fully simultaneous game; and (ii) the follower's profit exceeds the simultaneous play profit, while the leader's does not.

Consider now the case where the low-quality firm acts as the price leader, which is perhaps hardly justifiable on both theoretical and empirical grounds, but is nevertheless needed to complete the picture. Equilibrium qualities and profits are:

$$q_H^f(\overline{\theta}) = \overline{\theta} q_H^f(1) = 0.39999\overline{\theta};\quad q_L^l(\overline{\theta}) = \overline{\theta} q_L^l(1) = 0.19999\overline{\theta}; \tag{4.37}$$

$$\pi_H^f(\overline{\theta}) = \overline{\theta}^3 \pi_H^f(1) = 0.018\overline{\theta}^3;\quad \pi_L^l(\overline{\theta}) = \overline{\theta}^3 \pi_L^l(1) = 0.01199\overline{\theta}^3. \tag{4.38}$$

Observe that, as compared to the fully simultaneous equilibrium, (i) the high quality decreases, while the low quality increases; and (ii) again, as in the previous case, the leader is worse off than under simultaneous play.

An illustration is given in figure 4.1, where the cases of simultaneous play and sequential play with the low-quality leading are described. In order not to hinder the explanatory power of the figure, the remaining case where the high-quality firm takes the lead is not illustrated. The reaction functions pertaining to simultaneous play are represented by thick lines, while those describing the setting where the low-quality firm is leading are thin. The same applies to isoprofit curves. As quality levels change according to the specific timing chosen by firms in the price stage, the positions of reaction

for the fully simultaneous setting, and is omitted here. The same holds for the proof of uniqueness.

functions as well as the overall map of isoprofit curves change as well. Specifically, notice that the reaction functions of both firms move upwards as each firm (i) moves at the same time as the rival; (ii) moves earlier than the rival; (iii) moves later than the rival.

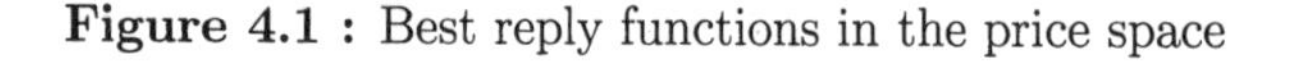
Figure 4.1 : Best reply functions in the price space

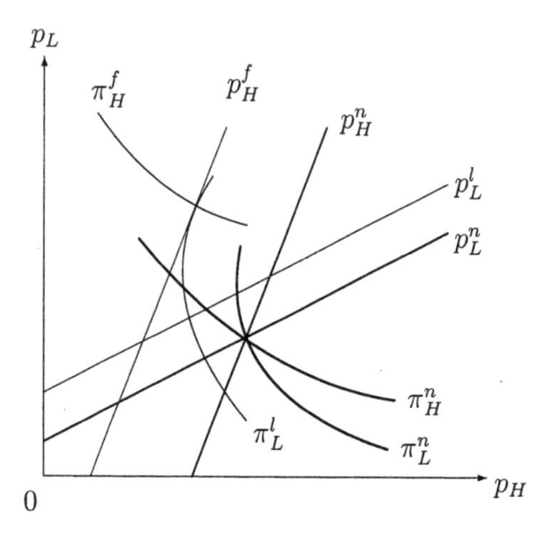

I am now in a position to consider the possibility that firms choose the timing of moves pertaining to the price stage before setting qualities in the first stage. The outcome of such a game is summarized by the following:

Proposition 4.4 *If firms can set the timing of moves in the price stage before deciding their respective quality levels, and have a commitment device, both will choose to move late, so that simultaneous play emerges.*

Proof. Since the size of $\overline{\theta}$ exerts only a scale effect on profits, I can confine to the case of $\overline{\theta} = 10$. Then, the game can be described by matrix 4.2.

		L	
		F	S
H	F	16.4; 12.1	15.06; 14.12
	S	18; 11.99	16.4; 12.1

Matrix 4.2

A quick inspection of matrix 4.2 suffices to verify that $\pi_i^f > \pi_i^N > \pi_i^l$, so that playing late (S) is a dominant strategy for both firms, and simultaneous play emerges at equilibrium, the latter being $\Omega^* = \{S_H(q_H^*, q_L^*), S_L(q_H^*, q_L^*)\}$.
■

4.5.2 Variable costs of quality improvement: Cournot competition

Turn now to the case where the downstream stage takes the form of competition in output levels. On the basis of the discussion carried out in the previous section, this setting can be quickly dealt with. The optimal qualities chosen when firms compete simultaneously in the output stage are:

$$q_H^n(\overline{\theta}) = \overline{\theta} q_H^n(1) = 0.369648\overline{\theta}; \quad q_L^n(\overline{\theta}) = \overline{\theta} q_L^n(1) = 0.292788\overline{\theta}, \tag{4.39}$$

yielding

$$\pi_H^n(\overline{\theta}) = \overline{\theta}^3 \pi_H^n(1) = 0.0176282\overline{\theta}^3; \quad \pi_L^n(\overline{\theta}) = \overline{\theta}^3 \pi_L^n(1) = 0.017478\overline{\theta}^3. \tag{4.40}$$

When the high-quality firm takes the lead in the market stage, the relevant equilibrium magnitudes are:

$$q_H^l(\overline{\theta}) = \overline{\theta} q_H^l(1) = 0.353321\overline{\theta}; \quad q_L^f(\overline{\theta}) = \overline{\theta} q_L^f(1) = 0.228453\overline{\theta}; \tag{4.41}$$

$$\pi_H^l(\overline{\theta}) = \overline{\theta}^3 \pi_H^l(1) = 0.0205598\overline{\theta}^3; \quad \pi_L^f(\overline{\theta}) = \overline{\theta}^3 \pi_L^f(1) = 0.0130477\overline{\theta}^3. \tag{4.42}$$

Finally, when the low-quality firm is appointed the leadership in the market stage, one obtains:

$$q_H^f(\overline{\theta}) = \overline{\theta} q_H^f(1) = 0.422087\overline{\theta}; \quad q_L^l(\overline{\theta}) = \overline{\theta} q_L^l(1) = 0.315747\overline{\theta}; \tag{4.43}$$

$$\pi_H^f(\overline{\theta}) = \overline{\theta}^3 \pi_H^f(1) = 0.0123216\overline{\theta}^3; \quad \pi_L^l(\overline{\theta}) = \overline{\theta}^3 \pi_L^l(1) = 0.0197048\overline{\theta}^3. \tag{4.44}$$

I am now in a position to state

Proposition 4.5 *If firms can set the timing of moves in the quantity stage before deciding their respective quality levels, and have a commitment device, both will choose to move early, so that simultaneous play emerges.*

Proof. Again, the size of $\overline{\theta}$ exerts only a scale effect on profits. Hence, I confine to the case of $\overline{\theta} = 10$. Then, the game can be described in reduced form by matrix 4.3.

		L	
		F	*S*
H	*F*	17.63; 17.48	20.6; 13.05
	S	12.32; 19.70	17.63; 17.48

Matrix 4.3

A quick inspection of matrix 4.3 reveals that $\pi_i^l > \pi_i^n > \pi_i^f$, so that playing early (F) is a dominant strategy for both firms. As a consequence, simultaneous play emerges at equilibrium, the latter being $\Omega^* = \{F_H(q_H^*, q_L^*), F_L(q_H^*, q_L^*)\}$. ■

Again, it is worth noting that, in the jargon of d'Aspremont and Gérard-Varet (1980, pp. 204-207), the quantity game is *strictly competitive*, i.e., it is not Stackelberg-solvable.

4.5.3 Fixed costs of quality improvement: Bertrand competition

Turn now to the case of a technology involving variable costs. From (4.28-4.29), it emerges that the choice of qualities affects the intercept of the reaction function of the high-quality firm in the price subgame, while it affects the slope of the low-quality firm's reaction function, in a way which recalls what we observed in the previous subsection. I can now look for the equilibrium qualities at the first stage. The FOCs of this problem are (cf. Motta, 1993, p. 116):

$$\frac{\partial \pi_H}{\partial q_H} = \frac{4\overline{\theta}^2 q_H(4q_H^2 - 3q_H q_L + 2q_L^2)}{(4q_H - q_L)^3} - 2q_H = 0; \tag{4.45}$$

$$\frac{\partial \pi_L}{\partial q_L} = \frac{\overline{\theta}^2 q_H^2(4q_H - 7q_L)}{(4q_H - q_L)^3} - 2q_L = 0. \tag{4.46}$$

Manipulating appropriately (4.45-4.46), yields the following equilibrium quality levels:[9]

$$q_H^n(\overline{\theta}) = 0.126650\overline{\theta}^2; \quad q_L^n(\overline{\theta}) = 0.024120\overline{\theta}^2, \tag{4.47}$$

[9] From an inspection of (4.46), it can be noticed that, if costs were nil, the low quality would be $q_L^n = 4q_H^n/7$. See Choi and Shin (1992).

where $q_H^n(1) = 0.12665$ and $q_L^n(1) = 0.02412$ are the qualities selected when $\overline{\theta}$ is equal to one. The corresponding equilibrium profits are:

$$\pi_H^n = 0.01222\overline{\theta}^4; \quad \pi_L^n = 0.0007640\overline{\theta}^4. \tag{4.48}$$

Again, this applies independently of the order of moves at the price stage. I turn now to the setting where each firm is alternatively appointed the leadership in the price stage. When the high-quality firm acts as the price leader, equilibrium qualities and profits are:

$$q_H^l(\overline{\theta}) = 0.12715\overline{\theta}^2; \quad q_L^f(\overline{\theta}) = 0.02949\overline{\theta}^2, \tag{4.49}$$

$$\pi_H^l = 0.01145\overline{\theta}^4; \quad \pi_L^f = 0.0009420\overline{\theta}^4. \tag{4.50}$$

It can be quickly verified that (i) both qualities increase as compared to the fully simultaneous game; and (ii) the follower's profit exceeds the simultaneous play profit, while the leader's is lower than that associated with simultaneous play.

Finally, the case where the low-quality firm is the price leader remains to be described. The equilibrium levels of qualities and profits are:

$$q_H^f(\overline{\theta}) = 0.12613\overline{\theta}^2; \quad q_L^l(\overline{\theta}) = 0.02425\overline{\theta}^2, \tag{4.51}$$

$$\pi_H^f = 0.01234\overline{\theta}^4; \quad \pi_L^l = 0.0007660\overline{\theta}^4. \tag{4.52}$$

Here, (i) the high quality decreases whereas the low quality increases as compared to the fully simultaneous game; and (ii) both the follower's and the leader's profits are higher than under simultaneous play.

Assume firms can decide the timing of their respective moves at the price stage before setting qualities in the first stage. The outcome of such an extended game is described by the following:

Proposition 4.6 *If firms can set the timing of moves in the price stage before deciding their respective quality levels, and have a commitment device, the high-quality firm will choose to move late whereas the low-quality firm will choose to move first, so that a unique equilibrium in pure strategies exists, involving sequential play.*

Proof. Again, provided that the size of $\overline{\theta}$ exerts only a scale effect on profits, I can confine to the case where $\overline{\theta} = 10$. Then, the game is described by matrix 4.4, which reveals that playing late (S) is a strictly dominant strategy for the high-quality firm.

		L	
		F	S
H	F	122.2; 7.64	114.5; 9.42
	S	123.4; 7.66	122.2; 7.64

Matrix 4.4

As a consequence, it is optimal for the low-quality firm to play early (F), and the unique equilibrium of this game, identified by the combination of strategies (S-F) involves sequential play with the low-quality firm in the price leader's position. Hence, in this game, $\Omega^* = \{S_H(q_H^*, q_L^*), F_L(q_H^*, q_L^*)\}$. ■

4.6 Time consistency

I am now in a position to discuss the issue of time consistency and the related role of commitment in the extended games with observable delay described above. In the first, firms' timing choices depend solely on the slope of reaction functions in the market stage or, equivalently, they are taken on the basis of a matrix game where profits are defined in terms of a generic quality pair. With both variable and fixed costs of quality improvement, price competition in the downstream stage leads firms to declare that they will move sequentially. Once they actually reach the market place and set prices, none of them has any incentive to renege the announcement made in the extension taking place between stages. This setting exactly replicates the situation depicted by HS (1990, Theorem III). Analogous considerations hold when market interaction takes the form of a Cournot game. Facing downward sloping reaction functions, firms find it optimal to move simultaneously at the earliest occasion (HS, 1990, Theorem II). When it comes to the second extended game with observable delay, where the extension is relocated upstream, before any decision in real time, the picture changes, and the credibility of announcements in some cases relies drastically on the existence of a commitment device. In the case of variable costs of quality improvements, the subgame perfect equilibrium involves simultaneous moves at the price stage, so that $\Omega^* \cap \overline{\Omega} = \emptyset$, and I can state

Remark 4.1 *The extended game with observable delay where firms bear variable costs of quality improvement and compete in prices is time inconsistent.*

Under Cournot competition, it appears that locating the timing decision ahead of the two stages taking place in real time or between them is irrelevant, in that $\Omega^* \equiv \overline{\Omega}$. Hence,

Remark 4.2 *The extended game with observable delay where firms bear variable costs of quality improvement and compete in quantities is strictly time consistent.*

In the case of fixed costs of quality improvements the subgame perfect equilibrium is unique and entails a particular sequential play with the low-quality firm leading. In other words, in the former setting the upward relocation of the timing choice yields an equilibrium which is not in the set of equilibria arising from the first extended game, while in the latter setting the relocation of the timing choice shrinks the set of equilibria to a single component of the wider set of equilibria yielded by the first extended game, i.e., $\Omega^* \subset \overline{\Omega}$. As a result, we have

Remark 4.3 *The extended game with observable delay where firms bear fixed costs of quality improvement and compete in prices is weakly time consistent.*

This discussion finally leads to the following

Proposition 4.7 *A sufficient condition for an extended game with observable delay to be strictly time consistent is that the timing choice immediately precede the stage which the choice itself refers to.*

Hence, the picture emerging from the above analysis highlights that the choice of timing in a game where firms (or players) choose more than one variable potentially gives rise to a problem largely analogous to that spotted by Amir (1995). When a generic quality pair is considered, profit functions are single-valued, so that HS's Theorems II-V hold. However, strategic interaction in the first stage generates different quality pairs according to the specific sequence of moves adopted in the market stage, so that when Stackelberg and Nash payoffs are evaluated from the viewpoint offered by the root of the two-stage game, each player's profit (or payoff) may or may not be monotone in the rival's action. If it is not, then timing decisions are inconsistent, i.e., HS's Theorem V may fail to apply.

A last issue remains to be investigated, namely, what happens if firms can endogenously and noncooperatively decide whether to plug the extension pertaining to the choice of timing at the root of the basic two-stage game, or to insert it between stages. This amounts to asking whether firms choose to play a time consistent game or not, or whether they prefer to set the timing at different points along the game tree. The case of Cournot competition is straightforward, in that any firm would always declare to move early irrespective of the location of the extension concerning itself as well as the rival. Hence, focus on the two Bertrand settings proposed above. In the case where production involves variable costs, the reduced form of the game in which firms can endogenously establish the position of the extension is given by matrix 4.5.

		L	
		B	R
H	B	16.53; 13.05	15.06; 14.12
	R	18; 11.99	16.4; 12.1

Matrix 4.5

Strategies B and R stand for *between* and *root*, respectively. The payoffs corresponding to (B, B) are those yielded by the correlated equilibrium. In the asymmetric cases where firm i set the timing of its price move at the root, while player j chooses between stages, player i declares it will move late, since at that stage qualities are already set and it becomes optimal to play a Stackelberg equilibrium. Observe that, since strategy R is strictly dominant for both firms, the equilibrium is (R, R), entailing that firms would choose to set the extension at the root, i.e., they would play a time inconsistent game due to a prisoners' dilemma.

		L	
		B	R
H	B	118.95; 8.54	114.5; 9.42
	R	123.4; 7.66	123.4; 7.66

Matrix 4.6

Turn now to the fixed cost setting. The reduced form of the game is in matrix 4.6, where obviously (R, B) and (R, R) yield the same profits. Again, R is a dominant strategy for both firms, strictly for H and weakly for L, so that firms choose to plug the extension at the root. As a result, they choose to play a weakly time consistent game. A final remark is in order. It appears from the analysis of matrices 4.5 and 4.6, as well as from the Cournot game which has not been explicitly investigated, that the taxonomy of the games in terms of time (in)consistency arising when both firms set the timing at the same point along the game tree carries over to the more general setting where the location of each player's declaration on timing is fully endogenous.

4.7 Concluding remarks

In this paper I have analysed the nature of the equilibria that can be expected to arise in extended duopoly two-stage games where firms first set the timing of moves pertaining to the market stage of the game, and then proceed to play. This may be the case when firms set variables that are bound to heavily affect the ensuing market competition, such as the amount of R&D effort, product quality or location.

I have obtained three main results. First, I have shown that the criteria for equilibrium selection introduced by Hamilton and Slutsky (1990) hold even without the requirement that the leader's profit be at least as high as in the simultaneous equilibrium. Indeed, this must be true in order for sequential play to arise as a subgame perfect equilibrium of the extended game, but it must hold for the follower as well. This leads to the second result. I have established that sequential play will be observed if and only if both firms are at least weakly better off playing sequentially than playing simultaneously, i.e., if the game is Stackelberg-solvable (d'Aspremont and Gérard-Varet, 1980; Robson, 1989). Third, resorting to a model of endogenous differentiation followed by price competition, I have proved the existence of cases where the leader can indeed be worse off than under simultaneous play. Finally, I have discussed the issue of time consistency in timing games, showing that a sufficient condition for such a choice to be strictly time consistent is that the timing stage be located adjacent to the stage at which firms will indeed be required to implement their timing decisions. Otherwise, as in Amir (1995), each player's payoff (or profit) function may not be monotone in the other player's choice, and HS's conclusions may not hold. The complete endogenization of the choice of timing has highlighted that firms can be expected to locate the extension in such a way that the resulting game may not be strictly time consistent.

Hence, HS's analytical framework is applicable to one-stage games where the choice of timing concerning the relevant strategic variable is not affected by any other strategic consideration. In the light of the foregoing analysis, it appears that in multi-stage games the location of the timing decisions along the tree becomes crucial.

The considerable range of models in which competition takes place in more than one stage suggests that the preferences over the distribution of roles and, consequently, the particular role distribution characterizing each specific model at equilibrium are issues to be carefully analysed in future research. Finally, the choice of timing could be extended to all the stages along which a game unravels (for applications to games with endogenous product differentiation, see Lambertini, 1996, 1997). A fully-fledged approach to this problem would certainly shed some new light on the explanatory value of the game-theoretic approach as to the behaviour of firms in the real world.

Appendices

Appendix 4.1: Equilibrium prices under sequential play and variable production costs

i) The equilibrium prices when the high-quality firm is the price leader are $p_H^l = q_H[2\overline{\theta}(q_H - q_L) + 2q_H^2 - q_H q_L + q_L^2]/[2(2q_H - q_L)]$ and $p_H^f = q_L[2\overline{\theta}(q_H - q_L) + 2q_H^2 + 3q_H q_L - q_L^2]/[4(2q_H - q_L)]$.

ii) The equilibrium prices when the low-quality firm is the price leader are $p_H^f = q_H[\overline{\theta}(q_H - q_L) + q_H^2 - q_H q_L + q_L^2]/(2q_H - q_L)$ and $p_H^l = [\overline{\theta} q_L(q_H - q_L) + q_H^2 q_L + 2q_H q_L^2 - q_L^3]/[2(2q_H - q_L)]$.

Appendix 4.2: Equilibrium outputs under sequential play and variable production costs

i) The equilibrium outputs when the high-quality firm is the quantity leader are $x_H^l = (2\overline{\theta} q_H - 2q_H^2 - \overline{\theta} q_L + q_L^2)/[2(2q_H - q_L)]$ and $x_L^f = (2\overline{\theta} q_H - 2q_H^2 - \overline{\theta} q_L - 4q_H q_L + q_L^2)/[4(2q_H - q_L)]$.

ii) The equilibrium outputs when the low-quality firm is the quantity leader are $x_H^f = (4\overline{\theta} q_H - 4q_H^2 - 3\overline{\theta} q_L + q_H q_L + 2q_L^2)/[4(2q_H - q_L)]$ and $x_L^l = q_H(\overline{\theta} + q_H - 2q_L)/[2(2q_H - q_L)]$.

Appendix 4.3: Equilibrium prices under sequential play and fixed production costs

i) The equilibrium prices when the high-quality firm acts as the price leader are $p_H^l = \overline{\theta} q_H(q_H - q_L)/(2q_H - q_L)$ and $p_L^f = \overline{\theta} q_L(q_H - q_L)/[2(q_H - q_L)]$.

ii) The equilibrium prices when the low-quality firm acts as the price leader are $p_H^f = \overline{\theta}(4q_H - q_L)(q_H - q_L)/[4(2q_H - q_L)]$ and $p_L^l = \overline{\theta}q_L(q_H - q_L)/[2(2q_H - q_L)]$.

References

1] Albæk, S. (1990), "Stackelberg Leadership as a Natural Solution under Cost Uncertainty", *Journal of Industrial Economics*, **38**, 335-47.
2] Amir, R. (1995), "Endogenous Timing in Two-Player Games: A Counterexample", *Games and Economic Behavior*, **9**, 234-37.
3] Battigalli, P. (1997), "Dynamic Consistency and Imperfect Recall", *Games and Economic Behavior*, **20**, 31-50.
4] Boyer, M. and M. Moreaux (1987a), "Being a Leader or a Follower: Reflections on the Distribution of Roles in Duopoly", *International Journal of Industrial Organization*, **5**, 175-92.
5] Boyer, M. and M. Moreaux (1987b), "On Stackelberg Equilibria with Differentiated Products: The Critical Role of the Strategy Space", *Journal of Industrial Economics*, **36**, 217-30.
6] Bulow, J., J. Geanakoplos and P. Klemperer (1985), "Multimarket Oligopoly: Strategic Substitutes and Complements", *Journal of Political Economy*, **93**, 488-511.
7] Choi, J.C. and H.S. Shin (1992), "A Comment on a Model of Vertical Product Differentiation", *Journal of Industrial Economics*, **40**, 229-31.
8] d'Aspremont, C. and L.-A. Gérard-Varet (1980), "Stackelberg-Solvable Games and Pre-Play Communication", *Journal of Economic Theory*, **23**, 201-17.
9] Davidson, C. and R. Deneckere (1986), "Long-Run Competition in Capacity, Short-Run Competition in Prices, and the Cournot Model", *RAND Journal of Economics*, **17**, 404-15.
10] Delbono, F., V. Denicolò and C. Scarpa (1996), "Quality Choice in a Vertically Differentiated Mixed Duopoly", *Economic Notes*, **25**, 33-46.
11] Dowrick, S. (1986), "von Stackelberg and Cournot Duopoly: Choosing Roles", *RAND Journal of Economics*, **17**, 251-60.
12] Fellner, W. (1949), *Competition among the Few*, New York, Kelly.
13] Gabszewicz, J.J. and J.-F. Thisse (1979), "Price Competition, Quality, and Income Disparity", *Journal of Economic Theory*, **20**, 340-59.
14] Gal-Or, E. (1985), "First Mover and Second Mover Advantages", *International Economic Review*, **26**, 649-53.

15] Hamilton, J.H. and S.M. Slutsky (1990), "Endogenous Timing in Duopoly Games: Stackelberg or Cournot Equilibria", *Games and Economic Behavior*, **2**, 29-46.

16] Kreps, D. and J. Scheinkman (1983), "Quantity Precommitment and Bertrand Competition Yield Cournot Outcomes", *Bell Journal of Economics*, **14**, 326-37.

17] Kydland, F. and E. Prescott (1977), "Rules rather than Discretion: The Inconsistency of Optimal Plans", *Journal of Political Economy*, **84**, 473-91.

18] Lambertini, L. (1996), "Choosing Roles in a Duopoly for Endogenously Differentiated Products", *Australian Economic Papers*, **35**, 205-24.

19] Lambertini, L. (1997), "Unicity of the Equilibrium in the Unconstrained Hotelling Model", *Regional Science and Urban Economics*, **27**, 785-98.

20] Lambertini, L. (1999), "Endogenous Timing and the Choice of Quality in a Vertically Differentiated Duopoly", *Research in Economics* (*Ricerche Economiche*), **53**, 101-09.

21] Lehmann-Grube, U. (1997), "Strategic Choice of Quality when Quality is Costly: The Persistence of the High-Quality Advantage", *RAND Journal of Economics*, **28**, 372-84.

22] Mailath, G.J. (1993), "Endogenous Sequencing of Firm Decisions", *Journal of Economic Theory*, **59**, 169-82.

23] Matsumura, T. (1999), "Quantity-Setting Oligopoly with Endogenous Sequencing", *International Journal of Industrial Organization*, **17**, 289-98.

24] Motta, M. (1993), "Endogenous Quality Choice: Price vs Quantity Competition", *Journal of Industrial Economics*, **41**, 113-31.

25] Ono, Y. (1982), "Price Leadership: A Theoretical Analysis", *Economica*, **49**, 11-20.

26] Pal, D. (1996), "Endogenous Stackelberg Equilibria with Identical Firms", *Games and Economic Behavior*, **12**, 81-94.

27] Piccione, M. and A. Rubinstein (1997a), "On the Interpretation of Decision Problems with Imperfect Recall", *Games and Economic Behavior*, **20**, 3-24.

28] Piccione, M. and A. Rubinstein (1997b), "The Absent-Minded Driver's Paradox: Synthesis and Responses", *Games and Economic Behavior*, **20**, 121-30.

29] Robson, A.J. (1989), "On the Uniqueness of Endogenous Strategic Timing", *Canadian Journal of Economics*, **22**, 917-21.

30] Robson, A.J. (1990a), "Duopoly with Endogenous Strategic Timing: Stackelberg Regained", *International Economic Review*, **31**, 263-74.

31] Robson, A.J. (1990b), "Stackelberg and Marshall", *American Economic Review*, **80**, 69-82.

32] Shaked, A. and J. Sutton (1982), "Relaxing Price Competition through Product Differentiation", *Review of Economic Studies*, **49**, 3-13.
33] Shaked, A. and J. Sutton (1983), "Natural Oligopolies", *Econometrica*, **51**, 1469-83.
34] Stackelberg, H. von (1934), *Marktform und Gleichgewicht*, Berlin, Springer.
35] Tirole, J. (1988), *The Theory of Industrial Organization*, Cambridge, MA, MIT Press.

Chapter 5

Existence of equilibrium in a vertically differentiated duopoly

Giulio Ecchia and Luca Lambertini[1]

5.1 Introduction

In the existing literature on vertical product differentiation, quality improvements imply alternatively a fixed or a variable cost. The nature of technology largely affects the equilibrium market structure (for a review, see Anderson *et al.*, 1992, chapter 8). The well known *finiteness property* obtains when quality improvements require either a fixed cost possibly represented by R&D efforts, or a variable cost which does not increase too fast as quality increases (Gabszewicz and Thisse, 1979, 1980; Shaked and Sutton, 1982, 1983). Otherwise, with sufficiently convex variable costs of quality, a segmented market structure obtains, as in horizontal differentiation models *à la* Hotelling (1929). As stressed by Gabszewicz and Thisse (1986), vertical differentiation models are generally expected to generate pure-strategy equilibria where prices are strictly above marginal production costs. On the contrary, under horizontal product differentiation, an established result is that a pure-strategy equilibrium in prices may not always exist (see, *inter alia*, d'Aspremont *et al.*, 1979;

[1]We thank Simon Anderson, Vincenzo Denicolò, Rudolf Kerschbamer and seminar audience in Bologna, Vienna, the XXV EARIE Conference (Copenhagen, August 1998) and the 1998 ASSET Conference (Bologna) for useful comments and discussion. Funding from the project *Regulation and quality in public services* (University of Bologna) is gratefully acknowledged. The usual disclaimer applies.

Gabszewicz and Thisse, 1986; Economides, 1986; Anderson, 1988). More precisely, a subgame perfect equilibrium with prices greater than marginal cost may fail to exist, because firms' location choices drive prices to marginal cost.[2] General results on existence and characterization of equilibria in markets for either horizontally or vertically differentiated products are derived by Anderson *et al.* (1997) under full market coverage and log-concave consumer distributions.

To our knowledge, all existing contributions on vertical product differentiation assume either partial or full market coverage. The only paper where the extent of market coverage is endogenously determined by firms' strategic interaction is due to Wauthy (1996), analysing a vertically differentiated duopoly where firms produce at no cost (as in Tirole, 1988; and Choi and Shin, 1992). Wauthy identifies the parameter ranges where either full or partial market coverage arises at equilibrium, as well as a range where a corner solution at the price stage obtains, in which the low-quality firm's price extracts all the surplus from the individual located at the lower bound of the support of consumer's distribution. He proves that such a corner solution is indeed the pure-strategy subgame perfect price equilibrium in the relevant range.

We consider a duopoly model of vertical differentiation with quadratic costs of quality improvements, so that the finiteness property does not hold. We investigate the existence and characterization of pure-strategy subgame perfect equilibria for a fixed market size. The alternative cases of full and partial market coverage are considered. We show that the parameter intervals in which the two alternative regimes can arise are disjoint. In order to define the demand structure in the parameter range where neither partial nor full market coverage can be properly defined, we prove that the low-quality firm sets its price to extract all the surplus of the poorest consumer in the market. Our findings reveal that, in such interval of demand parameters, a pure-strategy equilibrium fails to exist. This is due to the incentive for the high-quality firm to set a quality such that the rival's sales are driven to zero. Should firms produce at zero cost, as in Wauthy (1996), the non-existence problem would disappear, due to the incentive for the high-quality producer to supply the highest quality which is technologically feasible.

The remainder of the paper is structured as follows. The model is laid out in section 5.2, describing the alternative cases of partial and full market coverage. Section 5.3 contains the proof of the non-existence of a pure-strategy equilibrium with prices above marginal costs. Concluding remarks

[2] Obviously, a price equilibrium in mixed strategies always exists (Dasgupta and Maskin, 1986; Osborne and Pitchick, 1987).

are presented in section 5.4.

5.2 The model

We describe a model of vertically differentiated duopoly under complete information. Each firm produces a vertically differentiated good, with $q_H \geq q_L$, and then competes in prices against the rival. There exists a continuum of consumers indexed by their marginal willingness to pay for quality $\theta \in [\underline{\theta}, \overline{\theta}]$, with $\underline{\theta} = \overline{\theta} - 1$. The distribution of consumers is uniform, with density $f(\theta) = 1$, so that the total mass of consumers is also 1. Each consumer buys one unit of the product i that yields the highest net surplus $U = \theta q_i - p_i$, $i = H, L$.

Production technology involves variable costs, which are quadratic in the quality level and linear in the output level:

$$C_i = cq_i^2 x_i \,, \quad i = H, L, \; c > 0, \tag{5.1}$$

where x_i indicates the output level of firm i. Firm i's profit function is

$$\pi_i = (p_i - cq_i^2)x_i. \tag{5.2}$$

Competition between firms is fully noncooperative and takes place in two stages. In the first, firms set their respective quality levels; then, in the second, which is the proper market stage, they compete in prices. The solution concept applied is the subgame perfect equilibrium by backward induction. In the remainder of the section, we describe the two alternative equilibria that can arise under either full or partial market coverage.

5.2.1 Partial market coverage

Consider the case where the market is partially covered, i.e., there exists a consumer who is indifferent between buying the low-quality good and not buying at all. His location along the spectrum of the marginal willingness to pay is given by the ratio $k = p_L/q_L$. Given generic prices and qualities, the location of the consumer indifferent between the two varieties is $h = (p_H - p_L)/(q_H - q_L)$, so that now market demands are $x_H = \overline{\theta} - h$ and $x_L = h - k$. Given the cost function (5.1), the profit function of firm i remains defined as in (5.2).

Again, proceeding backwards, the FOCs for noncooperative profit maximization are

$$\frac{\partial \pi_H}{\partial p_H} = \overline{\theta} - \frac{2p_H - p_L - cq_H^2}{q_H - q_L} = 0; \tag{5.3}$$

$$\frac{\partial \pi_L}{\partial p_L} = \frac{p_H q_L - 2p_L q_H + c q_H q_L^2}{q_L(q_H - q_L)} = 0. \tag{5.4}$$

Observe that (5.3) coincides with (5.6) since the demand function for the high-quality good is the same in both settings. Solving the system (5.3-5.4), one obtains the following equilibrium prices:

$$p_H = \frac{q_H(2\overline{\theta} q_H + 2cq_H^2 - 2\overline{\theta} q_L + cq_L^2)}{4q_H - q_L}; \quad p_L = \frac{q_L(\overline{\theta} q_H + cq_H^2 - \overline{\theta} q_L + 2cq_H q_L)}{4q_H - q_L} \tag{5.5}$$

as the equilibrium prices. Solving the first stage, we obtain the Nash equilibrium qualities, $q_H^* = 0.40976\overline{\theta}/c$ and $q_L^* = 0.199361\overline{\theta}/c$.[3] Equilibrium prices are $p_H^* = 0.2267\overline{\theta}^2/c$; $p_L^* = 0.075\overline{\theta}^2/c$, outputs are $x_H^* = 0.2792\overline{\theta}$; $x_L^* = 0.3445\overline{\theta}$, while profits amount to $\pi_H^* = 0.0164\overline{\theta}^3/c$; $\pi_L^* = 0.0121\overline{\theta}^3/c$.

The equilibrium values of firms' profits are acceptable if total equilibrium demand is at most equal to one, i.e., $k \geq \underline{\theta}$, which implies the constraint $\overline{\theta} \leq 1.6032$. Otherwise, the marginal willingness to pay of the consumer supposedly indifferent between buying the low-quality good and not buying at all falls below the lower bound of the interval assumed for θ. If this is the case, i.e., $\overline{\theta} > 1.6032$, then the above specification of demand functions is no longer valid.

5.2.2 Full market coverage with an interior solution in prices

This setting follows the analysis presented in several contributions (Moorthy, 1988; Champsaur and Rochet, 1989; Cremer and Thisse, 1994; Lambertini, 1996). Suppose all consumers are able to buy, i.e., θ_1 is sufficiently high to allow for full market coverage. Given generic prices and qualities, the "location" of the consumer indifferent between the two varieties is $h = (p_H - p_L)/(q_H - q_L)$, so that market demands are $x_H = \overline{\theta} - h$ and $x_L = h - (\overline{\theta} - 1)$.

Consider the market stage. From the first order conditions (FOCs henceforth),

$$\frac{\partial \pi_H}{\partial p_H} = \overline{\theta} - \frac{2p_H - p_L - cq_H^2}{q_H - q_L} = 0; \tag{5.6}$$

[3] This can be verified through numerical calculations, initially performed by normalizing $\overline{\theta}$ to 1. Then, increasing $\overline{\theta}$ shows that the relationship between equilibrium qualities and $\overline{\theta}$ is linear. The proof that leapfrogging is not profitable is omitted since it is in Motta (1993), where a complete characterisation of this setting is presented.

$$\frac{\partial \pi_L}{\partial p_L} = \frac{p_H - 2p_L + cq_L^2}{q_H - q_L} - (\overline{\theta} - 1) = 0, \tag{5.7}$$

the following equilibrium prices obtain:

$$p_H = \frac{(q_H - q_L)(\overline{\theta} + 1) + 2cq_H^2 + cq_L^2}{3}; \; p_L = \frac{(q_H - q_L)(2 - \overline{\theta}) + 2cq_L^2 + cq_H^2}{3} \tag{5.8}$$

Substituting and rearranging, we get the profit functions defined exclusively in terms of qualities, $\pi_i(q_H, q_L)$. The subgame perfect quality levels are

$$q_H = \frac{4\overline{\theta} + 1}{8c}; \quad q_L = \frac{4\overline{\theta} - 5}{8c}, \tag{5.9}$$

which entails the general constraint $\overline{\theta} \geq 9/4$, in order for the poorest consumer to be in a position to buy the low-quality product. The corresponding equilibrium profits are $\pi_H = \pi_L = 3/(16c)$, and equilibrium demands are $x_H = x_L = 1/2$. Observe that the socially optimal qualities would be the first and third quartiles of the interval $[\underline{\theta}/2, \overline{\theta}/2]$, which obtains from the calculation of the preferred varieties for the richest and the poorest consumer in the market, if such varieties were sold at marginal cost. This implies that (i) qualities are set, respectively, too low and too high as compared to the social optimum; and (ii) this model shares its general features with the model of spatial competition with quadratic transportation costs (see Cremer and Thisse, 1994).[4]

5.2.3 Full coverage with a corner solution in prices

The demand functions are defined for full coverage, i.e., $x_H = \overline{\theta} - h$ and $x_L = h - \left(\overline{\theta} - 1\right)$, and we examine the case where p_H is endogenously determined by the first order condition for maximum profit, while p_L is chosen so as to extract the entire surplus from the poorest consumer:

$$p_L = \left(\overline{\theta} - 1\right) q_L \, . \tag{5.10}$$

From (5.6), we obtain the best reply of firm H :

$$p_H \left(p_L\right) = \frac{p_L + \overline{\theta} \left(q_H - q_L\right) + cq_H^2}{2} \tag{5.11}$$

[4]It can be shown that, under full market coverage, the spatial model with quadratic transportation costs is actually a special case of a vertical differentiation model with quadratic costs of quality improvement (see Neven, 1986; Cremer and Thisse, 1991).

which, in correspondence of (5.10), yields $p_H^* = \left(\overline{\theta} q_H - q_L + c q_H^2\right)/2$. The profit functions at the first stage are:

$$\pi_H = \frac{\left(\overline{\theta} q_H - q_L - c q_H^2\right)}{4\left(q_H - q_L\right)}; \pi_L = \frac{q_L\left(\overline{\theta} - 1 - c q_L\right)\left[q_H\left(2 - \overline{\theta} - c q_H\right) - q_L\right]}{2\left(q_H - q_L\right)}. \tag{5.12}$$

The FOC of firm H is:

$$\frac{\partial \pi_H}{\partial q_H} = \frac{\left(\overline{\theta} q_H - q_L - c q_H^2\right)\left[\overline{\theta}\left(q_H - 2q_L\right) + q_L\left(1 + 4c q_H\right) - 3c q_H^2\right]}{4\left(q_H - q_L\right)^2} = 0 \tag{5.13}$$

yielding four solutions, of which only one qualifies as the best reply function of firm H :

$$q_H\left(q_L\right) = \frac{\overline{\theta} - \sqrt{\overline{\theta}^2 - 4c q_L}}{2c} \tag{5.14}$$

since it's the only one satisfying the second order condition, which, in correspondence of (5.14) writes as follows:

$$\frac{\partial^2 \pi_H}{\partial q_H^2} = \Phi\left[\Omega + \Psi\right] \tag{5.15}$$

where:

$$\Phi = \frac{2c}{\left(\sqrt{\overline{\theta}^2 - 4c q_L} - \overline{\theta} + 2c q_L\right)^3} > 0; \ \Omega = \overline{\theta}\left(\sqrt{\left(\overline{\theta}^2 - 4c q_L\right)^3} - \overline{\theta}^3\right); \tag{5.16}$$

$$\Psi = c q_L\left[8c^2 q_L^2 - 2c q_L\left(\overline{\theta} + 2\right)^2 + 2\overline{\theta}^2\left(\overline{\theta} + 3\right) - 2\sqrt{\left(\overline{\theta}^2 - 4c q_L\right)^3}\right]. \tag{5.17}$$

Given that $\Phi > 0$, $\partial^2 \pi_H / \partial q_H^2 \propto \Omega + \Psi \in \mathbb{R}$ for all $q_L \in \left[0, \overline{\theta}^2/(4c)\right]$. Solving $\Omega + \Psi = 0$ we obtain:

$$q_L = 0; \ q_L = \frac{\overline{\theta}^2}{4c} \tag{5.18}$$

with $\Omega + \Psi < 0$ for all $q_L \in \left(0, \overline{\theta}^2/(4c)\right)$. Expression (5.14) is indeed a best reply iff:

$$q_H\left(q_L\right) = \frac{\overline{\theta} - \sqrt{\overline{\theta}^2 - 4c q_L}}{2c} \geq q_L \Leftrightarrow q_L \geq \frac{\overline{\theta} - 1}{c}. \tag{5.19}$$

However, observe that, in correspondence of (5.14), the profits of firm L are:

$$\pi_L = q_L\left(\overline{\theta} - 1 - cq_L\right) > 0 \text{ for all } q_L < \frac{\overline{\theta} - 1}{c}. \tag{5.20}$$

Moreover, under (5.14), we have $x_H = \pi_H = 0$. The foregoing discussion can be summarised as follows:

A] If $q_L \geq \left(\overline{\theta} - 1\right)/c$, then $q_H\left(q_L\right) \geq q_L$; $\pi_H = 0$ and $\pi_L \leq 0$.

B] If $q_L < \left(\overline{\theta} - 1\right)/c$, then $q_H\left(q_L\right) < q_L$; $\pi_H = 0$ and $\pi_L > 0$.

Case [A] cannot be a duopoly equilibrium because neither firm obtains positive profits; in particular, firm L will exit since, in general, its profits are negative. Case [B] is not an equilibrium outcome because it entails leapfrogging.

The discussion carried out in this section leads to the following:

Proposition 5.1 *If* $\overline{\theta} \in [1, 1.6032]$, *there exists a unique subgame perfect equilibrium in pure strategies, yielding partial market coverage. If* $\overline{\theta} \geq 9/4$, *there exists a unique subgame perfect equilibrium in pure strategies, yielding full market coverage. If* $\overline{\theta} \in (1.6032, 9/4)$, *the duopoly equilibrium does not exist in pure strategies.*

5.3 Concluding remarks

In the foregoing analysis, we have investigated the existence of a pure-strategy subgame perfect equilibrium in a duopoly model of vertical differentiation with convex variable costs of quality. We have shown that there are parameter ranges where a pure strategy equilibrium exists (i) under partial market coverage, if consumers' marginal willingness to pay for quality is relatively low; (ii) under full market coverage, if consumers' marginal willingness to pay for quality is relatively high. However, these two intervals are disjoint. In such intermediate parameter range, we have proved that the low-quality firm is constrained to price so as to extract all the surplus from the poorest consumer in the market. This, in turn, induces the high-quality firm to decrease its quality towards the rival's, in order to increase its market share. This argument, in combination with the possibility for the low-quality firm to leapfrog the rival, entails that a pure-strategy duopoly equilibrium does not exist. This cannot happen in a model where production costs are nil, as assumed by Wauthy (1996).

The above findings reveal that, contrary to previous beliefs, vertical differentiation models suffer from a problem of non-existence of the equilibrium in pure strategies which affects spatial differentiation models. While in spatial models the non-existence is due to an insufficient degree of convexity of transportation costs, in vertical differentiation models it appears to be due to the convexity of production costs in a subset of the parameter space where a corner solution in prices is the unique candidate as a Nash equilibrium at the market stage.

References

1] Anderson, S.P. (1988), "Equilibrium Existence in the Linear Model of Spatial Competition", *Economica*, **55**, 479-91.
2] Anderson, S.P., A. de Palma and J.-F. Thisse (1992), *Discrete Choice Theory of Product Differentiation*, Cambridge, MA, MIT Press.
3] Anderson, S.P., J.K. Goeree and R. Ramer (1997), "Location, Location, Location", *Journal of Economic Theory*, **77**, 102-27.
4] Champsaur, P. and J.-C. Rochet (1989), "Multiproduct Duopolists", *Econometrica*, **57**, 533-57.
5] Choi, J.C. and H.S. Shin (1992), "A Comment on a Model of Vertical Product Differentiation", *Journal of Industrial Economics*, **40**, 229-31.
6] Cremer, H. and J.-F. Thisse (1991), "Location Models of Horizontal Differentiation: A Special Case of Vertical Differentiation Models", *Journal of Industrial Economics*, **39**, 383-90.
7] Cremer, H. and J.-F. Thisse (1994), "Commodity Taxation in a Differentiated Oligopoly", *International Economic Review*, **35**, 613-33.
8] Dasgupta, P. and E. Maskin (1986), "The Existence of Equilibrium in Discontinuous Economic Games, II: Applications", *Review of Economic Studies*, **53**, 27-42.
9] d'Aspremont, C., J.J. Gabszewicz and J.-F. Thisse (1979), "On Hotelling's 'Stability in Competition'", *Econometrica*, **47**, 1145-50.
10] Economides, N. (1986), "Minimal and Maximal Differentiation in Hotelling's Duopoly", *Economics Letters*, **21**, 67-71.
11] Gabszewicz, J.J. and J.-F. Thisse (1979), "Price Competition, Quality and Income Disparities", *Journal of Economic Theory*, **20**, 340-59.
12] Gabszewicz, J.J. and J.-F. Thisse (1980), "Entry (and Exit) in a Differentiated Industry", *Journal of Economic Theory*, **22**, 327-38.
13] Gabszewicz, J.J. and J.-F. Thisse (1986), "On the Nature of Competition with Differentiated Products", *Economic Journal*, **96**, 160-72.

14] Hotelling, H. (1929), "Stability in Competition", *Economic Journal*, **39**, 41-57.
15] Lambertini, L. (1996), "Choosing Roles in a Duopoly for Endogenously Differentiated Products", *Australian Economic Papers*, **35**, 205-24.
16] Moorthy, K.S. (1988), "Product and Price Competition in a Duopoly Model", *Marketing Science*, **7**, 141-68.
17] Motta, M. (1993), "Endogenous Quality Choice: Price vs Quantity Competition", *Journal of Industrial Economics*, **41**, 113-32.
18] Neven, D. (1986), "Address Models of Differentiation", in Norman, G. (ed.), *Spatial Pricing and Differentiated Markets*, London, Pion.
19] Osborne, M.J. and C. Pitchick (1987), "Equilibrium in Hotelling's Model of Spatial Competition", *Econometrica*, **55**, 911-22.
20] Shaked, A. and J. Sutton (1982), "Relaxing Price Competition through Product Differentiation", *Review of Economic Studies*, **69**, 3-13.
21] Shaked, A. and J. Sutton (1983), "Natural Oligopolies", *Econometrica*, **51**, 1469-83.
22] Tirole, J. (1988), *The Theory of Industrial Organization*, Cambridge, MA, MIT Press.
23] Wauthy, X. (1996), "Quality Choice in Models of Vertical Differentiation", *Journal of Industrial Economics*, **44**, 345-53.

Chapter 6

MQS regulation and predatory behaviour

Luca Lambertini and Carlo Scarpa[1]

6.1 Introduction

The (limited) empirical evidence on the effects of minimum quality standards (MQS) is seemingly at odds with most of the theoretical results in duopoly, which typically point to the desirability of MQS and to a quality-improving effect of their introduction.[2] For instance, Carroll and Gaston (1991) - analysing the regulation of electricians - indicate that licensing restrictions reduces supply[3] and decreases average quality (increase accident rates). More recent work on this topic by Chipty and Witte (1997) points in the same direction, showing how minimum standards increase the probability that a firm leaves the market. Furthermore, they observe that the maximum quality supplied may decline after the introduction of the MQS, whenever the MQS induces some firm to leave the market.

The ideas that emerge from these papers is that the introduction of a MQS affects market structure, in particular inducing exit by some firms, and that this decrease in competition can be associated with quality reductions. In this paper we present a model where these effects are present, showing that

[1]Some of the points developed in this paper emerged from comments by Massimo Motta on previous work by one of the authors, and we are grateful for his contribution. We thank Pierpaolo Battigalli and Emanuela Carbonara for useful comments.

[2]This is true of most models considering duopolistic markets (starting from Ronnen, 1991). Scarpa (1998) obtains a different result with three firms, while Maxwell (1998) points out potential negative effects on the incentive to innovate.

[3]The supply reducing effect is confirmed, for instance, by Gormley (1991).

the introduction of a MQS can provide a basis for predatory behavior by the high quality firm, and that the likelihood that the MQS reduces supply is substantially higher than indicated, for instance, by the seminal paper by Ronnen (1991).

Previous literature in duopolistic markets, including Donnenfeld and Weber (1992 and 1995), Crampes and Hollander (1995), Constantatos and Perrakis (1997 and 1998), Ecchia and Lambertini (1997) and Lutz (1997), is typically based on two assumptions, which are unnecessarily restrictive and which we try to relax in the current paper.

The first one is that, following the introduction of the MQS, the high quality firm will play the duopoly equilibrium strategy, and will not try to drive the rival out of the market. This possibility clearly exists, as the introduction of an MQS limits the ability of the low quality firm to differentiate its output from the rival's. It is sometimes claimed that this seriously undermines the conclusions of previous models, but a formal analysis is still lacking.

Within this framework, we show that the duopoly survives only for very low levels of the MQS. For higher values, predatory behaviour is indeed possible, and within a certain interval two equilibria exist: the duopoly equilibrium and a "predatory" one, where the high quality firm reduces its quality level and the rival leaves the market. When both equilibria exist, we use the criterion of risk-dominance (Harsanyi and Selten, 1988) in order to select the "most likely" equilibrium. This criterion allows us to rule out the duopoly equilibrium for the entire admissible range of parameters.

The remainder of the paper is organised as follows. The setup is laid out in section 6.2. Predatory behaviour when qualities can be adjusted costlessly is described in section 6.3. Section 6.4 contains concluding remarks.

6.2 The model

We consider a vertically differentiated duopoly, where firms H and L produce goods of either high (H) or low (L) quality. If q denotes quality, firm i's cost is $C_i = C(q_i)$, with C', C'' > 0 for all $q_i \in [0, \infty)$, $i = H, L$.

A population of consumers indexed by a taste parameter θ is uniformly distributed over the interval $[0, \overline{\theta}]$.[4] The total mass of consumers is thus equal to $\overline{\theta}$. Consumers buy either one unit of one of the two goods or nothing. The representative consumer's utility function is $U = \max[\theta q - p; 0]$. Marginal consumers are identified by $\theta_H = (p_H - p_L)/(q_H - q_L)$ and $\theta_L = p_L/q_L$.

[4] In this section the assumption of uniformity of the distribution of consumers plays no role. Later on it will be important to explicitly solve the model.

Therefore, market demands are

$$x_H = \overline{\theta} - \frac{p_H - p_L}{q_H - q_L} \text{ and } x_L = \frac{p_H - p_L}{q_H - q_L} - \frac{p_L}{q_L} \tag{6.1}$$

and the profit function of firm i is $p_i x_i(q_i, q_j) - C(q_i)$.

We only analyse subgame perfect equilibria where firms first of all choose whether or not to enter and with what quality levels; at a second stage, firms decide prices. Notice that the decision to enter a market coincides with the decision on the quality level[5] and that in both stages choices are simultaneous. In an unregulated equilibrium, firms produce different quality levels in order to soften price competition (Gabszewicz and Thisse, 1979; Shaked and Sutton, 1982); we denote these levels by q_H^U and q_L^U. The effects of a minimum quality standard (s) have been studied by Ronnen (1991), who proves that a value S exists such that, if $s \in [q_L^U, S]$,

- an equilibrium exists, such that both firms survive (for $s > S$, the low quality firm leaves the market);

Furthermore, relative to the unregulated case:

- in this equilibrium, both quality levels are higher;
- social welfare is larger;
- the high quality seller's profit is lower;
- a value S' exists, such that if $s \leq S' < S$, the low quality seller's profit increases.[6]

If a MQS is introduced and both firms survive, firm i will produce quality $q_i(s) \geq s$.[7] Given that prices are chosen optimally, equilibrium profits will be

$$\pi_H = \frac{4\overline{\theta}^2 [q_H(s)]^2 (q_H(s) - q_L(s))}{(4q_H(s) - q_L(s))^2} - C[q_H(s)] \tag{6.2}$$

[5] Given that in this kind of model set-up costs are typically absent, only when a firm defines and starts producing a good can we say that it is "in a market".

[6] These results hold with Bertrand competition. With Cournot competition, Valletti (2000) shows that the optimal MQS is below the lowest unregulated quality, and is thus ineffective.

[7] If the standard is binding, the L firm can either produce a quality level equal to it, or else leapfrog its rival. Therefore, $q_L(s) \in \{s, q^{LF}(q_H(s))\}$, where $q^{LF}(q_H(s)) > q_H(s)$ is the quality level the L firm would choose to optimally leapfrog the rival.

$$\pi_L = \frac{\overline{\theta}^2 q_H(s)(q_H(s) - q_L(s))q_L(s)}{(4q_H(s) - q_L(s))^2} - C[q_L(s)] \tag{6.3}$$

If the low quality firm produces $q_L = s$ and the H firm plays its duopoly best reply $q_H^D(s)$, the profit levels will be denoted by π_H^D and π_L^D, respectively. This is the case usually analysed in the literature.

6.2.1 Equilibrium predation

Here we want to analyse one aspect that the literature typically does not consider, i.e. that there exist values of $s \leq S$ where a second equilibrium exists, in which the high quality firm forces its rival out of the market. In this equilibrium, as we will see, most of the above conclusions are reversed.

As the reaction function of the low quality firm is not continuous in $q_L = s$, the high quality firm may find it profitable to adopt a predatory behaviour rather than the best reply $q_H^D(s)$. Indeed, a quality level lower than $q_H^D(s)$ might squeeze the low quality demand to a level that forces the low quality firm to leave the market.[8] On the other hand, if the H firm's quality level is too low, leapfrogging might be profitable, so that predation may no longer be feasible.[9]

Let us denote the predatory quality by $q_H^P(s)$, yielding a monopoly profit $\pi_M^P(q_H^P(s))$. Predation requires the L firm to have no option but to leave the market, while the H firm earns profits larger than it would in a duopoly equilibrium. Formally, the necessary and sufficient condition for predation is that the following inequalities are simultaneously satisfied:

$$\pi_L(q_H^P(s), q_L(s)) \leq 0 \tag{6.4}$$

and

$$\pi_M^P(q_H^P(s)) \geq \pi_H^D(q_H^D(s), s) \tag{6.5}$$

The first condition simply states that when $q_H = q_H^P(s)$ there is no level of q_L such that the L firm can make positive profits: thus, given predation, the optimal choice of firm L is to leave the market. Condition (6.5) indicates that predation must be at least as profitable to the high quality firm as the duopoly equilibrium. When this condition is satisfied, predation is indeed optimal for the H firm.

---

[8] Remember that in the relevant range of parameters quality levels are strategic complements, so that reaction functions are increasing (see, e.g., Motta, 1993).

[9] The term *leapfrogging* refers to a situation in which the firm which used to produce the lower quality good reacts to the quality choice of its rival by setting a quality level higher than q_H (see Motta *et al.*, 1997).

Is predation a Nash equilibrium? The question is legitimate, as when firm L leaves the market, sticking to a predatory quality would not be optimal any longer for firm H; however, as the decision to enter the market coincides with the decision to produce a certain quality level, changing the quality level is not an option when we consider predation.[10] Indeed, as the only cost we consider - in line with the literature on quality competition and MQS - is associated with quality, nothing in the model indicates that entry (or exit) is less irreversible than quality choices. As firms are already present in the market, firm L must simply decide between (a) leaving the market ($q_L = 0$) or (b) staying with a certain quality level $q_L > 0$.

Define $Q^P \equiv \{q_H \mid \pi_L(q_H(s), q_L(s)) \leq 0 \text{ and } \pi_M^P(q_H(s)) \geq \pi_H^D(q_H^D(s), s)\}$. This is the set of all predatory quality levels. Two remarks are in order.

- The set Q^P is an interval, whose length depends on s. Notice that, depending on the value of s, the monopoly quality level without entry threats (q^M) may or may not belong to Q^P.
- As predation requires a relatively low quality level, unless $q^M \in Q^P$ predation requires the firm to distort its choice: when $q_H = q_H^P \neq q^M$, if the L firm leaves the market $\partial \pi_H / \partial q_H > 0$. This entails a cost for the firm, increasing in the difference between the predatory quality level and q^M.

It is thus straightforward to prove that

Lemma 6.1 *If $q^M \in Q^P$, then $q_H^P = q^M$. If $q^M \notin Q^P$, then $q_H^P = \sup Q^P$.*

As long as $q_L = s$, predation is preferable to accepting the duopoly whenever (6.5) is satisfied. Hence, for some admissible values of s, we may have two equilibria. It is possible to show that predation is easier, the larger the value of s.

Proposition 6.1 *Given $s \leq S$, if $q_H^P(s)$ exists, then $dq_H^P(s)/ds \geq 0$.*

Proof. As long as $q^M \in Q^P$, the predatory quality does not vary with s. Thus, let us consider the case $q^M \notin Q^P$.

Whenever the MQS is binding, $\partial \pi_L / \partial s < 0$, for a given level of q_H.[11] On the other hand, given q_L, from (6.3) it can be verified that $\partial \pi_L / \partial q_H > 0$.

[10] On the general logic of predatory behaviour, see Ordover and Saloner (1989). Notice that one fairly straightforward alternative would be to re-state the model as a model of entry barrier, where the MQS can be used to preempt rivals'entry.

[11] Notice that - as shown by Ronnen (1991) - this does not hold when q_H is *not* given.

Notice that $q_H^P(s)$ is defined as the value such that $\pi_L(q_H^P(s), s) = 0$. Applying the implicit function theorem, we can conclude that in this case

$$\frac{dq_H^P}{ds} = -\frac{\partial \pi_L / \partial s}{\partial \pi_L / \partial q_H} > 0. \tag{6.6}$$

This proves the claim. ■

When the MQS increases, the space available to the low quality firm shrinks, and thus predation can be achieved with a quality level closer to q_M. This is clearly preferable for the high quality producer, in that its best reply $q_H^D(s)$ is larger than $q_H^P(s)$. Any decrease in q_H relative to the best reply entails a "cost", so that predation is indeed made "cheaper" by higher levels of s.

Proposition 6.1 implies that predation - whenever feasible - may be profitable either for all values within $[0, S]$, or in a connected subset of this interval.[12]

On the other hand, if the H firm decreases its quality level "too much", leapfrogging may be feasible. Denote the profit of the L firm which leapfrogs the predator as $\pi^{LF} \equiv \pi(q_H^P; q^{LF}(q_H^P))$. Then, we derive the following:

Proposition 6.2 *If $\pi_M^P(q_H^P) < \pi_H^D$, then the unique equilibrium is $(q_H^D(s); s)$.*

If $\pi_M^P(q_H^P) > \pi_H^D$, and $\pi^{LF} > 0$, then again the unique equilibrium is $(q_H^D(s); s)$.

If $\pi_M^P(q_H^P) > \pi_H^D$ and $\pi^{LF} < 0$, then two equilibria exist; the first one is, again, $(q_H^D(s); s)$, while the second one is $(q_H^P(s); s)$.

Moreover:

Proposition 6.3 *There exist values of the standard $\underline{q}$ and $\overline{q}$, with $\underline{q} \leq \overline{q} \leq S$, such that*

(i) for $s < \underline{q}$ when $q_L = s$ predation is not profitable;

(ii) for $s \in [\underline{q}; \overline{q})$ if $q_H = q_H^P$ the best reply of the other firm is to leapfrog $(\pi^{LF}(q_H^P; q^{LF}(q_H^P)) > 0)$;

(iii) for $s \in [\overline{q}; S]$ predation is profitable.

[12] If one only considers a situation where both firms survive, a standard below unregulated quality levels does not affect market equilibrium. However, in principle, when we consider predation, a MQS might matter even if $s < q_L^U$.

Proof. Obvious, along the lines of the Proof of Proposition 1 (in particular, given property (6.6)). ■

Therefore, as shown in figure 6.1, the interval $[q_L^U, S]$ can be divided in three regimes indicated in this Proposition. In the third one, when predation is profitable we have the aforementioned multiplicity and thus we have a potential problem of equilibrium selection, which may be solved with the use of the notion of risk-dominance.[13]

Figure 6.1 : Equilibria for different levels of s

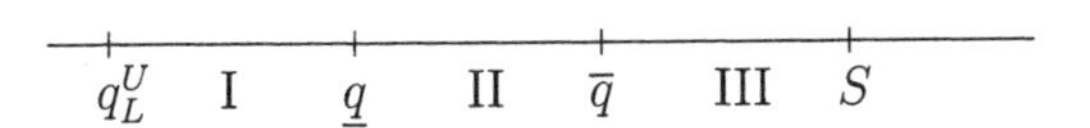

Interval I. Duopoly
Interval II. Duopoly (under threat of leapfrogging)
Interval III. Two equilibria: (a) Duopoly (under threat of leapfrogging);
(b) Predation

At the final stage, prices are always chosen optimally given quality levels. At the first stage, the strategy space of each firm is binary, with $q_H \in \{q_H^D(s), q_H^P(s)\}$ and $q_L \in \{s, 0\}$ where $q_L = 0$ implies that the low quality firm exits the market. The reduced form of the game is described by Matrix 6.1.

		L	
		s	0
H	$q_H^D(s)$	π_H^D, π_L^D	$\pi_M^D, 0$
	$q_H^P(s)$	π_H^P, π_L^P	$\pi_M^P, 0$

Matrix 6.1

[13]See Harsanyi and Selten (1988). The existing plethora of equilibrium concepts potentially opens numerous - too many - possibilities. The choice of risk-dominance as a selection criterion is due first of all to the relation between risk-dominance and trembling-hand perfection (Kajii and Morris, 1997), a criterion that in the situation examined seems intuitively appropriate. Notice that an analogous choice was done by Motta *et al.* (1997).

The two Nash equilibria of this game are the duopoly equilibrium ($q_H^D(s)$, s) and the predatory equilibrium ($q_H^P(s)$, 0), while π_M^D and π_i^P are out-of-equilibrium payoffs. π_M^D is the monopoly profit that the high quality firm obtains when it produces the best reply to the MQS and the rival exits the market. π_i^P is the duopoly profit for firm i ($i = H, L$) when the high quality firm produces the predatory quality and the low quality firm sticks to the MQS.

Applying the criterion of risk-dominance to this case we can see that the predatory equilibrium risk-dominates the other one when

$$(\pi_H^D - \pi_H^P)\pi_L^D > -(\pi_M^P - \pi_M^D)\pi_L^P \tag{6.7}$$

This inequality is a complex expression in s, which may or may not hold within the framework adopted so far. We can investigate this aspect and the actual feasibility of predatory strategies by slightly specialising the model relative to Ronnen.

6.3 Predation in the standard model

Let us consider the model of the previous section, setting $\overline{\theta} = 1$ and $C(q_i) = q_i^2$.[14] Unregulated quality levels are $q_L^U = 0.024119$ and $q_H^U = 0.126655$. As to the behaviour of the regulator in choosing the quality standard s, the following holds:

Remark 6.1 *If the high quality firm plays the best response to s, the market can remain a duopoly if and only if $s \leq 0.04809$.*

Proof. Solving the game backwards, suppose firms play the Nash equilibrium prices. The best reply function of the high quality firm is

$$q_H(q_L) = \frac{6q_L + 1}{24} + \frac{1 - 6q_L}{24\left(1 - 9q_L + 324q_L^2 + 3q_L\sqrt{3(23 - 208q_L + 3888q_L^2)}\right)^{1/3}}$$
$$+ \frac{\left(1 - 9q_L + 324q_L^2 + 3q_L\sqrt{3(23 - 208q_L + 3888q_L^2)}\right)^{1/3}}{24} \tag{6.8}$$

[14] This is just meant to facilitate the derivation of explicit results. As $\overline{\theta}$ increases, it can be shown that $q_i(\overline{\theta}) = \overline{\theta}^2 q_i(\overline{\theta} = 1)$ and $\pi_i(\overline{\theta}) = \overline{\theta}^4 \pi_i(\overline{\theta} = 1)$. For a proof, see Motta (1993).

Plugging (6.8) into (6.3), we can see that $\pi_L(q_H^D(q_L), q_L) = 0$ at $q_L = 0.04809$. This establishes that $S = 0.04809$. ■

Define the social welfare function as the sum of profits and consumer surplus, $SW = \sum_i \pi_i + CS$, where

$$CS = \int_{\theta_L}^{\theta_H} (\theta q_L - p_L) d\theta + \int_{\theta_H}^{1} (\theta q_H - p_H) d\theta. \tag{6.9}$$

Plugging (6.8) into the welfare function, it is easily verified that social welfare is concave for all $q_L \in [0.02412, 0.06761)$. Therefore, it is easy to show that, absent predation, it is socially optimal to set $s = 0.04809$.

As for the possibility of predation, a preliminary step consists in calculating the monopoly optimum (without threats). From (6.1), the demand function is $x = 1 - p/q$, where p/q is the generic price-quality ratio offered by the monopolist. Accordingly, the monopolist's profit function is $\pi^M = px - q^2$. Given the monopoly price $p^M = q/2$, the profit function simplifies to $\pi^M = q(1/4 - q)$, which is maximised at $q^M = 1/8$.

Consider now predatory behaviour by the high-quality firm. The condition $\pi_L = 0$ yields

$$\hat{q}_H = \frac{q_L\left[(8q_L - 1) - \sqrt{1 - 12q_L}\right]}{2(16q_L - 1)}. \tag{6.10}$$

As stated in lemma 6.1, whenever $q^M \leq \widehat{q}_H$, then $q_H^P = q^M$. Otherwise, given a generic MQS s, this defines the predatory quality level: $q_H^P = \hat{q}_H$. Numerical calculations show that $q^M = \widehat{q}_H$ at $q_L = 0.047366$; therefore, we have established that, without leapfrogging,

$$q_H^P = \frac{s\left[(8s - 1) - \sqrt{1 - 12s}\right]}{2(16s - 1)} \;\; \forall\, s \in (0.02412, 0.04737]; \tag{6.11}$$

$$q_H^P = q^M = \frac{1}{8} \;\; \forall\, s \in [0.04737, 0.04809]. \tag{6.12}$$

It is fairly obvious that monopoly profits are larger than $\pi_H^D(s)$. This indicates that predation is certainly feasible and profitable for all $s \in (0.04737, 0.04809]$. It remains to be established whether, in the remainder of the admissible range for q_L, the quality level defined by (6.11) is optimal, i.e. whether $\pi_M^P(q_H^P(s)) > \pi_H^D(q_H^D(s), s)$. Numerical calculation reveals that

$$\pi_M^P > \pi_H^D(q_H^D(s), s) \quad \forall\, s \in (0.03356, 0.04737], \tag{6.13}$$

and conversely for all $s \in (0.02412, 0.03356]$.[15] Therefore:

Remark 6.2 *Whenever* $s \leq 0.03356$, *predation is not profitable.*

This indicates that for very low levels of s, predation would require the high quality firm to distort its choice by "too much", and would thus be too costly. In the previous notation, this means that $\underline{q} = 0.03356$.

So far we have assumed that $q_L = s$. Let us now focus on the possibility for the low-quality firm to leapfrog the rival, if the latter tries to predate.

With $q^{LF}(q_H^P) > q_H^P$, the profit from leapfrogging is:

$$\pi^{LF} = \frac{4[q^{LF}(q_H^P)]^2(q^{LF}(q_H^P) - q_H^P)}{(4q^{LF}(q_H^P) - q_H^P)^2} - [q^{LF}(q_H^P)]^2, \tag{6.14}$$

with $q^{LF}(q_H^P) > q_H^P$, $q_H^P = \min\{\hat{q}_H, q^M\}$. Evaluating (6.14) for all $q_H^P \in [0.054503, 1/8]$, reveals that

$$\pi^{LF}(q^{LF}(q_H^P); q_H^P) > 0 \quad \forall\, q_H^P \in [0.054503, 1/12). \tag{6.15}$$

Moreover, $q_H^P = 1/12$ when $s = 0.04117$. Therefore, we can state

Remark 6.3 *Given* $s \in (0.03356, 0.04117)$, *any* $q_H = q_H^P(s)$ *makes leapfrogging profitable.*

To sum up, a low level of the MQS ($s \leq 0.03356$) makes predation unprofitable even if the low quality firm sticks to $q_L = s$. A slightly higher level of s allows predation only if the low quality firm keeps producing the minimum quality level, but this is not rational. Indeed, in the interval indicated in remark 6.3, if the H firm assumes that the rival produces $q_L = s$ and in turn produces the correspondent "predatory" quality level, the (previously) low quality firm will not produce s, but will leapfrog the rival. Therefore, in this case predation will not take place in equilibrium. This leads to the final conclusion

Proposition 6.4 *Whenever* $s \in [0.04117, 0.04809]$ *an equilibrium exists, in which the high quality firm produces* $q_H = q_H^P(s)$, *forcing the rival to exit the market.*

[15] When $s = 0.033556$, the corresponding predatory quality is $q_H^P = 0.054503$.

When the minimum quality standard is set within this interval, the high quality firm will produce q_H^P and the low quality firm will not be able to obtain positive profits either producing $q_L = s$ or leapfrogging the rival. Given that the high quality firm adopts a predatory behaviour, the market is too small for two firms to coexist.

Notice that, in the interval of proposition 6.4, a second equilibrium exists, in which $q_L = s$ and $q_H = q_H^D(s)$. In order to select one equilibrium, we can use the criterion of risk dominance, already introduced in (6.7). Plugging expressions (6.8), (6.10), (6.12) and (6.11) into (6.7), we obtain an extremely complex expression in s, which can be evaluated numerically to find that (6.7) is satisfied for all $s \in [0.04117; 0.04809]$. This establishes the final result of this section:

Proposition 6.5 *Whenever two equilibria exist, the criterion of risk dominance selects the equilibrium with predation.*

This result is hardly surprising, given the analogy between risk dominance and trembling-hand perfection (see Kajii and Morris, 1997). If the H firm produces the predatory quality and the L firm insists on supplying the minimum quality standard, the latter firm is going to suffer substantial losses, while its potential gain (the duopoly profit π_L^D) is anyway sufficiently close to zero.

We are now in a position to evaluate the regulator's perspective. Given the duopolistic structure, welfare increases with s as long as $s < 0.04809$. For future reference, define this welfare function as $SW_s^D(s)$.[16] On the basis of propositions 6.4 and 6.5, predation gives rise to a monopoly for $s \in [0.04117, 0.04809]$. However, we are going to show that this may be a lesser evil from a social standpoint. Indeed, given the optimal monopoly price schedule $p^M = q/2$, the associated socially optimal MQS is $s^M = 3/16$. Equilibrium demand is $x_s^M = 1/2$, and the resulting welfare level is $SW_s^M = 9/256$.

Now we can establish that $SW_s^D(s) > 9/256$ for all $s \in (0.02893, 0.04117)$, and conversely for all s outside this interval. Therefore, given this result and Propositions 4 and 5, we have:

Proposition 6.6 *Consider the admissible range of s in duopoly, i.e., $s \leq 0.04809$, and compare the regulated duopoly equilibrium with the second-best monopoly where $s^M = 3/16$. The regulator's preferred market structure is the following:*

(i) a second-best monopoly for all $s \in [0.024119, 0.02893]$;

[16] We omit the expression of social welfare in the regulated duopoly, for the sake of brevity. It can be easily calculated using (6.9) and the firms' profit functions.

(ii) a duopoly for all $s \in (0.02893, 0.04117]$;

(iii) a second-best monopoly for all $s > 0.04117$.

Maximum welfare is achieved in duopoly at $s = 0.04117$.

The situation is represented in figure 6.2 (notice that the scale is arbitrary). Recall that the optimal MQS ($s = 0.04117$) allows the duopoly to survive, but leaves the low quality firm with zero profit.

Figure 6.2 : Social welfare

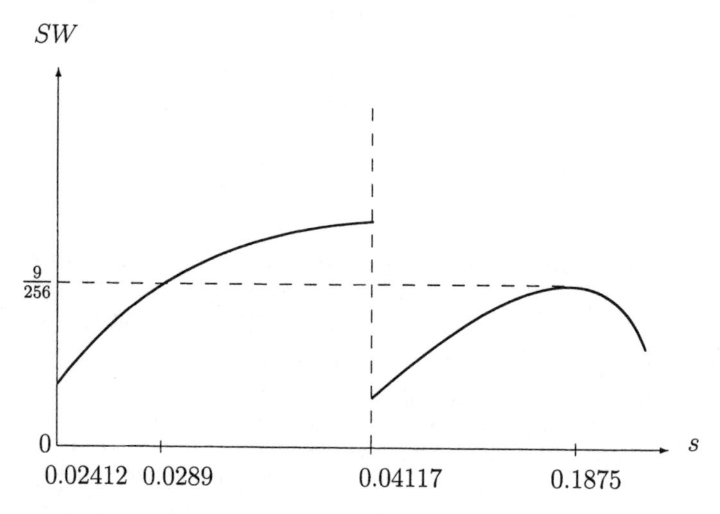

6.4 Concluding remarks

While the initial literature on minimum quality standards tended to stress the positive welfare effects of such regulatory interventions, this paper, in line with a growing theoretical and empirical literature, points in the opposite direction. The introduction of an MQS can have repercussions on market structure, opening the possibility of predatory behaviour. Moreover, whenever predation is an equilibrium, it is selected by the risk dominance criterion.

Although some of the basic assumptions of the standard model we have adopted look quite restrictive, we do not feel that any of our results are particularly model-specific, as long as Bertrand competition is considered.[17]

One possible extension consists in considering the adjustment costs associated with the unanticipated introduction of the standard. We feel this is potentially relevant as the structure of the multi-stage game logically requires the quality choice to be a long term choice, that constrains the firm for a significant period of time. Therefore, models without adjustment costs can hardly provide a coherent representation of how incumbent firms *react* to the new standard.[18] However, it can be shown that this does not modify substantially the above conclusions.[19] Probably, the most relevant extension of the present analysis should consider a variable cost technology as in Crampes and Hollander (1995), but we leave this for future research.

References

1] Carroll, S. and R. Gaston (1991), "Occupational Restrictions and the Quality of Service Received: Some Evidence", *Southern Economic Journal*, **47**, 959-77.

2] Chipty, T. and A. D. Witte (1997), "An Empirical Investigation of Firms' Reponses to Minimum Quality Standards Regulations", NBER Working Paper no. 6104.

3] Constantatos, C. and S. Perrakis (1997), "Vertical Differentiation: Entry and Market Coverage with Multiproduct Firms", *International Journal of Industrial Organization*, **16**, 81-103.

4] Constantatos, C. and S. Perrakis (1998), "Minimum Quality Standards, Entry, and the Timing of the Quality Decision", *Journal of Regulatory Economics*, **13**, 47-58.

[17]If firms optimize w.r.t. output levels, then the analysis carried out by Valletti (2000) holds.

[18]This does not mean that standard models are incorrect. Simply, they compare different, alternative scenarios (with and without MQS, or with MQS of different sizes) but not the *change* in behaviour of a firm and of an industry after the introduction of the MQS.

[19]We have solved the model under the assumption that costs are

$$C^a(q_i) = q_i^2 + (q_i - q_i^U)^2$$

where the second component on the r.h.s. is the adjustment costs. In case adjustment costs are asymmetric, i.e. if only *increasing* quality is costly, predatory behaviour is even more rewarding. Detailed calculations are available upon request.

5] Crampes, C. and A. Hollander (1995), "Duopoly and Quality Standards", *European Economic Review*, **39**, 71-82.
6] Donnenfeld, S. and S. Weber (1992), "Vertical Product Differentiation with Entry", *International Journal of Industrial Organization*, **10**, 449-72.
7] Donnenfeld, S. and S. Weber (1995), "Limit Qualities and Entry Deterrence", *RAND Journal of Economics*, **26**, 113-30.
8] Ecchia, G. and L. Lambertini (1997), "Minimum Quality Standards and Collusion", *Journal of Industrial Economics*, **45**, 101-13.
9] Gabszewicz, J.J. and J.-F. Thisse (1979), "Price Competition, Quality and Income Disparities", *Journal of Economic Theory*, **20**, 340-59.
10] Gormley, W. (1991), "State Regulations and the Availability of Child Care Services", *Journal of Policy Analysis and Management*, **10**, 78-95.
11] Harsanyi, J.C. and R. Selten (1988), *A General Theory of Equilibrium in Games*, Cambridge, MA, Harvard University Press.
12] Kajii, A. and S. Morris (1997), "The Robustness of Equilibria to Incomplete Information", *Econometrica*, **65**, 1283-1309.
13] Lutz, S. (1997), "Vertical Product Differentiation and Entry Deterrence", *Journal of Economics*, **65**, 79-102.
14] Maxwell, J. W. (1998), "Minimum Quality Standards as a Barrier to Innovation", *Economics Letters*, **58**(3), 355-360.
15] Motta, M. (1993), "Endogenous Quality Choice: Price vs Quantity Competition", *Journal of Industrial Economics*, **41**, 113-32.
16] Motta, M., Thisse J.-F. and A. Cabrales (1997), "On the Persistence of Leadership or Leapfrogging in International Trade", *International Economic Review*, **38**, 809-24.
17] Ordover, J. and G. Saloner (1989), "Predation, Monopolization, and Antitrust", in R. Schmalensee and R. Willig (eds) *Handbook of Industrial Organization*, Vol. 1, Amsterdam, North-Holland, 537-96.
18] Ronnen, U. (1991), "Minimum Quality Standards, Fixed Costs, and Competition", *RAND Journal of Economics*, **22**, 490-504.
19] Scarpa, C. (1998), "Minimum Quality Standards with More than Two Firms", *International Journal of Industrial Organization*, **16**, 665-76.
20] Shaked, A. and J. Sutton (1982), "Relaxing Price Competition through Product Differentiation", *Review of Economic Studies*, **49**, 3-13.
21] Valletti, T. (2000), "Minimum Quality Standards under Cournot Competition", *Journal of Regulatory Economics*, **18**, 235-45.

Chapter 7

Intraindustry trade and the choice of technology

Luca Lambertini and Gianpaolo Rossini

7.1 Introduction

Nowadays trade is made up, to a large extent, of vertically differentiated manufactured goods and services aimed to appease the tastes of consumers enjoying different degrees of affluence and willingness to pay. Quality differences change the nature of competition among firms and also comparative advantage whenever international trade is considered.

When goods of different quality enter the picture firms in oligopolistic markets enjoy a broader set of strategic decision variables with respect to the case of homogeneous goods. For instance, firms may set either prices or quantities while sticking to the manufacturing of goods of a particular standard.

Sometimes, the quality reputation reached by a firm in the eyes of buyers is the result of a long standing commitment coming from a previous stage of infant development when the relevant market was bound to be the domestic one. Here, a kind of protection or simply national bias would provide a shelter.

Alternatively, or in addition, the quality commitment of a firm may be the aftermath of a production choice undertaken in a market safeguarded either by some patented process or product whose protection is geographically limited to the home country or to the integrated area where the firm belongs.[1]

[1]There are many instances of commercial disputes among firms as to the geographic effectiveness of a patent. Sometimes, the result is a partial insulation of the home market

A further hot question related to the quality choice concerns the kind of competition that occurs both in protected markets and under free trade. Quality choice and market strategies are indeed strictly related and are affected by the technological menu to the avail of producers.

In spite of the complexity of the relationship between technology, quality and market strategies, causal evidence and literature in the field (Mussa and Rosen, 1978; Shaked and Sutton, 1982; 1983) are of great help and suggest two main routes. The first is associated with a technology that makes quality improvements achievable only bearing a variable cost. In the second case, quality calls for a fixed cost that can be interpreted as an R&D effort. This menu may be broadened, as we shall see, introducing different degrees of efficiency in each technology.

Literature, and evidence, provide instances of price (Shaked and Sutton, 1984; Lambertini, 1997; Motta, Thisse and Cabrales, 1997) and quantity competition (Motta, 1992), in both a short and a long-run perspective. In some cases qualities do not change when moving from a protected environment to free trade. In others the elimination of barriers stimulates the appearance of a fresh quality standard. These contributions examine the implications of free trade for domestic welfare and market structure when production opportunities are the same over the world, while countries differ in terms of income distribution, i.e., willingness to pay for quality.

Contributions on the effects of technological choices, i.e. variable versus fixed cost technology, are rather scanty and, therefore, worth examining. We shall see that quality competition assumes various shapes according to the technological environment in which it is cast. The choice of technology out of a menu, or the availability of a certain technology inherited from the past, determines the nature of competition. In such circumstances, welfare may be increased by trade policies coordinating quality standards, when firms fail to do so.

We offer two scenarios for the analysis.

In the *first*, firms have excess capacity and compete *à la* Bertrand. With technologies of heterogeneous efficiency, market incentives let firms shun co-

for a firm that has introduced either a new product or a new process in the country in which it has its main establishment. Then, a condition akin to autarchy arises. "For instance, relying on its patent rights, a Japanese exporter could, until very recently, prohibit the sale of lower-price parallel imports from the EC in Japan. Similarly, on the basis of its own patent rights, a competing European industry can still prevent lower-priced parallel imports of its products from Japan and other third countries in the EC...This is permitted by the WTO" (Bronckers, 1996, p.16). Other examples are recent cases of limitations of the validity of patents for some pharmaceuticals to the domestic market of the country in which they have been first produced.

ordination. Only through governments' intervention is a welfare-maximizing equilibrium attained. Implementation mimics a trade policy partially similar to that adopted by the EU Commission, i.e. a contingent tariff that is triggered when the price of an imported good in the EU is sufficiently lower than the price of the competing domestically produced good. "The European Commission, responsible for the injury [*due to presumed dumping*] investigation, seems to focus on the level of the foreign price undercutting....the extent to which the price of a foreign product in the European market is lower than the European price for a similar product" (Vandenbussche, 1996, p.116).[2] An antidumping duty can then be inflicted, for instance, on a good imported from South Korea or the US or Canada. Surprisingly enough, this tariff may have a procompetitive effect. In this sense it could even be operated by an antitrust agency concerned with welfare maximization.

The *second* scenario features firms which are endowed with a limited productive capacity in the short-medium run and then behaving *à la* Cournot. When these firms have the opportunity of choosing between quantity or quality as their strategic variable, different equilibria emerge, according to the affluence of customers. With a sufficiently low reservation price, *firms coordinate spontaneously over quality* adopting the very same technology. If the market is rich, making for a higher willingness to pay, multiple standards appear. However, it would be socially desirable to have firms converging to a unique quality standard adopting the most efficient technology.

Policies will have to be tuned to the consumers' evaluation of quality, on which welfare depends. If such evaluation is relatively low there is no need for either national or common international standards, since an efficient common standard is the endogenous outcome of market play. In other words, for low income countries trading with similar countries the coordination over quality standards is the automatic result of free trade and market competition.

The opposite happens when consumers exhibit high evaluation of quality. Then the variance of standards matters more for welfare while no endogenous automatic coordination obtains. In such circumstances, governments may set the quality standard that corresponds to a Pareto-superior equilibrium.[3]

[2] See also Vandenbussche (1995), Veugelers and Vandenbussche (1999), Bronkers (1996).

[3] The time consistency of these interventions is analyzed in Jensen and Thursby (1996). In an uncertain R&D race between firms of different countries, they show that a domestic (autarchic) standard is bound to be time inconsistent.

We shall not face time inconsistency, because it is optimal for governments to set the same standard regardless of the market regime.

There exists a wide literature (see Katz and Shapiro, 1985 and 1986; Farrell and Saloner, 1985; 1987; Matutes and Regibeau, 1988, *inter alia*), where the issue of standardization depends on compatibility among products, with or without network externalities.

The remainder of the chapter is organized as follows. In section 7.2 we illustrate the autarky model, while in section 7.3 we go through price competition mode of behaviour in free trade. Section 7.4 is dedicated to quantity competition. Section 7.5 provides concluding comments.

7.2 The basic setting

We first confine the analysis to *autarky* and consider a country with only one firm producing a differentiated good of quality q sold to a population of consumers with uniform income distribution, i.e. with one consumer for each level of income.

On the demand side, the marginal willingness to pay for the differentiated good is defined by $\theta \in [\underline{\theta}, \overline{\theta}]$ with $\underline{\theta} = \overline{\theta} - 1$ and $\underline{\theta} > 0$.[4] Population density is normalized to 1, making the total mass of consumers in the market equal to 1.

Each consumer has unit demand and buys if and only if his net surplus is non-negative, i.e. $U = \theta q - p \geq 0$, where p is the price charged by the monopolist. The market demand function is therefore $x = \overline{\theta} - p/q$.

On the supply side, we can consider two alternative types of technology. The *first* is given by the following cost function:

$$C_v = tq^2 x, \tag{7.1}$$

convex in quality and linear in quantity, which does not require any fixed cost. Parameter $t \in (0, 1]$ is for different degrees of efficiency.

The *second* technology looks as follows:

$$C_f = tq^2, \tag{7.2}$$

i.e. only a fixed cost is incurred.[5] The profit function is:

$$\pi_i = px - C_i \text{ for } i = f, v. \tag{7.3}$$

[4]This formalization of consumers' preferences has been introduced by Mussa and Rosen (1978). Since θ can be seen as the reciprocal of the marginal utility of nominal income, we can interpret the reservation price as the closest proxy to consumers' affluence (see Tirole, 1988, p. 96).

[5]For instance this can be the case in the software industry where firms bear high R&D costs, while average variable costs are negligible.

For $i = v$, defined as *variable cost technology*, optimality gives:[6]

$$q_v = \frac{\overline{\theta}}{3t}; \qquad p_v = \frac{2\overline{\theta}^2}{9t}; \quad x_v = \frac{\overline{\theta}}{3}; \quad \pi_v = \frac{\overline{\theta}^3}{27t}; \quad W_v = \frac{\overline{\theta}^3}{18t}; \tag{7.4}$$

where W stands for social welfare.

For $i = f$, defined as *fixed cost technology*, we get

$$q_f = \frac{\overline{\theta}^2}{8t}; \qquad p_f = \frac{\overline{\theta}^3}{16t}; \quad x_f = \frac{\overline{\theta}}{2}; \qquad \pi_f = \frac{\overline{\theta}^4}{64t}; \quad W_f = \frac{\overline{\theta}^4}{32t}. \tag{7.5}$$

While firms always prefer technology v to f,[7] social preferences switch from v to f when $\overline{\theta} > 16/9$. Consumers are partially served with both technologies if $\overline{\theta}$ is less than 2. We shall take into account this constraint on $\overline{\theta}$ in the remainder of the paper.

In the following sections we abandon autarky and deal with various kinds of competition in the international market. A common feature of all settings is the inheritance, from autarky or from a period of market insulation due either to trade barriers or home biases, of an irreversible choice relating either to capacity or to quality. Quality may be stuck to the autarky level by patents protecting acquired R&D knowledge.

7.3 Bertrand competition and free trade

The focus is on a free trade setting, with no barrier of any kind in a two-country world. First, firms compete in prices, with distinct technologies and different roles in the market game.

Price competition is a very common strategy adopted in mature and also advanced industries whenever capacity is not a crucial constraint. Despite the stern effect of price competition in a free trade environment we observe across countries price differentials not justified on the basis of existing barriers. As a result, most trade and competition authorities in advanced countries regard price differentials as a symptom of some kind of dumping. In this section we come across circumstances in which unilateral tariffs as a safeguard measure may be secured if price differentials show up conspicuously between domestic and foreign goods. These trade policies are socially desirable.

[6]Independently of the technology adopted, the monopolist supplies the same quality as the social planner, since the demand function is linear (Spence, 1975, pp. 419-21).

[7]Note that $\pi_v > \pi_f$ for all $\overline{\theta} < 64/27$, which is larger than 2. Since partial market coverage obtains under technology f for all $\overline{\theta} < 2$, this entails that firms prefer technology v.

7.3.1 Variable costs and free trade

Free trade takes place between two countries competing in an industry where each one has a single firm adopting the variable cost technology. The two countries are equal in all respects but for the efficiency of their respective firms. One firm has a cost function as in (7.1), while the other has the same cost function with $t = 1$. The quality of the good sold by the less efficient firm is lower than the one of the more efficient firm. The former firm is labeled L while the latter is called H. The same code is used for the two countries.

Quality standards are set in autarky as an irreversible commitment due to an arbitrarily large adjustment cost, possibly because of a patent. Demand functions result from the integrated market, whose density is twice as large as that of the previous autarchic market.

Technical argument. We identify the positions of the consumers indifferent between buying either of the two goods, as $h = (p_H - p_L)/(q_H - q_L)$, and between buying the low quality good or nothing, as $k = p_L/q_L$. Market demands are:

$$x_H = 2\left[\overline{\theta} - \frac{3t(p_H - p_L)}{\overline{\theta}(1-t)}\right]; \quad x_L = \frac{6(tp_H - p_L)}{\overline{\theta}(1-t)}. \tag{7.6}$$

Profit functions look as follows:

$$\pi_H = 2\left(p_H - \frac{\overline{\theta}^2}{9t}\right)\left[\overline{\theta} - \frac{3t(p_H - p_L)}{\overline{\theta}(1-t)}\right]; \ \pi_L = \frac{2(\overline{\theta}^2 - 9p_L)(tp_H - p_L)}{3\overline{\theta}(t-1)}. \tag{7.7}$$

Firms noncooperatively and simultaneously optimize w.r.t. prices, obtaining $p_H = \overline{\theta}^2(8-5t)/\left[9t(4-t)\right]; \ p_L = \overline{\theta}^2(2-t)/\left[3(4-t)\right]$, and then $p_H/p_L \geq 1$ for all $t \in (0,1]$. Substituting prices into (7.7) and rearranging, yields the Nash equilibrium profits

$$\pi_H^n = \frac{32\overline{\theta}^3(1-t)}{27t(t-4)^2}; \qquad \pi_L^n = \frac{8\overline{\theta}^3(1-t)}{27(t-4)^2}. \tag{7.8}$$

It is quickly established that $\pi_H^n > \pi_L^n$.■

With price competition between two firms based in two distinct countries and enjoying different levels of efficiency in quality supply, consumers in both countries reach the same level of surplus. However, the level of producer surplus turns out to be higher in the more efficient country H.

7.3.2 Endogenous choice of roles

The above results pertain to a scenario in which firms play in a simultaneous way, i.e. none has a first mover advantage. It is more interesting to see what happens when firms can move in a sequential way. Do firms choose to play sequentially, when we introduce a preplay stage where they noncooperatively set the timing? Following Hamilton and Slutsky (1990), we analyze an extended game with observable delay, consisting of two stages. In the first, firms have the option between playing at the first available occasion (F) or delay as long as possible (S). If both select the same strategy, a simultaneous equilibrium holds. Otherwise, a sequential equilibrium is singled out of the two possible. Then firms optimally set prices according to the timing of moves previously decided. The market equilibrium becomes part of the two-stage subgame perfect equilibrium of the extended game with observable delay.[8]

Technical argument. If firm H takes the lead, its program is:

$$\max_{p_H} \pi_H = 2\left(p_H - \frac{\overline{\theta}^2}{9t}\right)\left[\overline{\theta} - \frac{3t(p_H - p_L)}{\overline{\theta}(1-t)}\right] \tag{7.9}$$

$$\text{s.t.: } p_L = \frac{\overline{\theta}^2 + 9tp_H}{18}.$$

The leader's price is $p_H = \overline{\theta}^2(3t-4)/[9t(t-2)]$ and equilibrium profits are $\pi_H^l = 4\overline{\theta}^3(1-t)/[27t(2-t)]$ and $\pi_L^f = 2\overline{\theta}^3(1-t)/[27(2-t)^2]$, where superscripts l and f stand for leader and follower, respectively.

If firm L moves first and firm H follows, the leader's problem is:

$$\max_{p_L} \pi_L = \frac{2(\overline{\theta}^2 - 9p_L)(tp_H - p_L)}{3\overline{\theta}(t-1)} \tag{7.10}$$

$$\text{s.t.: } p_H = \frac{4\overline{\theta}^2 - 3\overline{\theta}^2 t + 9tp_L}{18t},$$

whose solution is $p_L = \overline{\theta}^2(3-2t)/[9(2-t)]$, and equilibrium profits are $\pi_L^l = \overline{\theta}^3(1-t)/[27(2-t)]$; $\pi_H^f = \overline{\theta}^3(t-4)^2(1-t)/[54t(t-2)^2]$. ■

The comparison between Stackelberg (or sequential Nash) and Nash (or simultaneous Nash) equilibrium profits gives rise to the following sequence of inequalities:

[8] For further technical details on extended games, see ch. 4.

$$\pi_i^f > \pi_i^l > \pi_i^n, \quad i = H, L, \tag{7.11}$$

allowing for a solution of the first stage of the game, as in matrix 7.1.

		H: F	H: S
L	F	$\pi_L^n; \pi_H^n$	$\pi_L^l; \pi_H^f$
	S	$\pi_L^f; \pi_H^l$	$\pi_L^n; \pi_H^n$

Matrix 7.1

The game has two asymmetric subgame Nash perfect equilibria involving sequential moves, (F, S) and (S, F). There exist no generally accepted criteria for selecting one of these equilibria which cannot be Pareto-ordered. However, aggregate industry profits provide some hints as to the desirability of one equilibrium *vis à vis* the other, from firms' viewpoint. By comparing the aggregate payoffs associated with (F, S) and (S, F), we observe that $\pi_L^f + \pi_H^l > \pi_H^f + \pi_L^l$ for all $t \in (0, 1]$.

Since the game has two asymmetric equilibria and industry profits are higher when the best firm takes the lead, the more efficient firm should take the lead of moves in the market as far as price setting is concerned. This will benefit both firms.

Yet, how socially desirable is the result preferred by firms? To answer, we discuss policy-makers' preferences by considering a similar game of timing, described in matrix 7.2, where payoffs are social welfare levels.

		H: F	H: S
L	F	$W_L^n; W_H^n$	$W_L^l; W_H^f$
	S	$W_L^f; W_H^l$	$W_L^n; W_H^n$

Matrix 7.2

Given that $W_i^n > W_i^l > W_i^f$, $i = H, L$, a trade policy, aimed at preventing the foreign firm from playing the follower's role, will lead to the unique Nash equilibrium (F, F), where firms set prices simultaneously at the first available occasion. The first best solution from a social standpoint could be reached if both countries simultaneously adopted a trade policy, that would entail a contingent tariff equal to the difference between the follower's equilibrium price and the simultaneous Nash equilibrium price for the same firm. Setting this tariff is not easy in practical terms since the two alternative equilibria are not observable. However, if the most efficient firm moves first there is room for a welfare improvement either via some kind of taxation on the laggard to push firms to set prices simultaneously or by obliging firms to publish prices at the same date. A tariff on the low quality firm may change the incentive to produce for the foreign market since the lower quality good is going to be sold at a gross price that equals the price of the high quality good. Therefore the low quality good is dominated and not purchased by consumers abroad. Then, the social desirability of simultaneous play may become the ground for coordination of trade policies towards firms producing goods of different quality standards and competing in prices.

If governments do so they benefit consumers in both countries, since they induce firms to play a non-cooperative Nash equilibrium in prices. Incidentally, this is what an antitrust agency would aim to do, even though this may appear quite odd, since tariffs and antitrust policies usually do not pursue the same objective and they often disprove each other. However, once we allow for coordination, a country's tariff against the foreign firm is equivalent to an antitrust policy towards the domestic firm. The observational equivalence between antitrust and trade policies in this setting is due to the existence of multiple subgame perfect equilibria in the extended game played by firms. If there were only one subgame perfect Stackelberg (or sequential Nash) equilibrium, the tariff policy would maintain its usual anticompetitive characterization. While, in this case, it is just the way to push firms to play the most competitive alternative strategy.

7.3.3 Heterogeneous technologies

Here firms utilise different technologies. Firms H and L may alternatively supply their respective qualities through $C_f = tq^2$ or $C_v = q^2x$, with $t \in (0,1]$. This setting produces several subcases where specific viability conditions on parameters $\overline{\theta}$ and t must be satisfied. Provided that such constraints are met, asymmetric technologies in the extended Bertrand game lead to the same subgame perfect equilibrium outcome, with sequential play.[9] Since both firms strictly prefer the follower's role, such equilibria cannot be Pareto-ordered and a case arises for trade policy. Again, a tariff on the imports from the least efficient firm leads to simultaneous play and benefits consumers of

[9] Detailed calculations are omitted for the sake of brevity. They are available upon request.

both countries. The measures we figure out are somehow close to those undertaken by the EU Commission, i.e., contingent trade policies triggered by relevant differences in prices of goods of the same industry.

As a conclusion to this section we state the following:

Proposition 7.1 *Under Bertrand competition, firms prefer sequential play irrespective of their technological endowments. Welfare maximizing governments may independently implement a trade policy having a procompetitive effect since it decreases the 'distance' among firms making them move in a simultaneous way. This implies removal of price leadership.*

7.4 Cournot competition and free trade

Cournot competition occurs if quantity is bounded by capacity constraints driving marginal cost to infinity when producing beyond threshold (Kreps and Scheinkman, 1983; Davidson and Deneckere, 1986).

Several new scenarios occur, where either quality or output levels are the legacy of autarky according to the features of firms' technologies embodied in their cost functions. With a fixed cost technology, firms can only adjust output. With a variable cost technology, they have the option to fix either output or quality at the autarky level.[10] Efficiency is the same for both technologies, i.e., $t = 1$. We do not reproduce the same technological constraints as in the previous section.[11]

7.4.1 Symmetric choices

(a) Quality competition under capacity constraints. Firms stick to the output of autarky, while being free in terms of quality supplied in duopoly because of the flexibility of variable cost technology. Equilibrium prices are driven by demand, as in Kreps and Scheinkman (1983), in that, if firms optimize in prices capacity bites. The resulting Bertrand equilibrium mimics the Cournot outcome. The capacity constrained equilibrium comes from either price setting, whereby firms are driven by demand to full capacity utilization, or quantity setting, with firms producing at full capacity and selling at market-clearing prices. The ensuing analysis is then a complement of unconstrained price setting in section 7.3.

[10]Observe that a Minimum Quality Standard policy is possible only if firms remain flexible in quality.

[11]The reproduction of the same cases can be provided upon request from the authors.

(b) Quantity competition under quality constraints. Here is the opposite scenario. Qualities are fixed in autarky. Firms compete in quantities.
(c) Cournot competition with fixed costs. When both firms operate with fixed cost technology, it is reasonable to assume that they can adjust only quantities under free trade, since qualities are the result of R&D investments undertaken in autarky, seen as a period of patent shelter. Qualities are as in (7.5), so that goods are perfect substitutes.

We sum up the above cases as follows:

Proposition 7.2 *Provided that technological choices are symmetric, both firms strictly prefer to be flexible in quality-setting rather than in quantity-setting, for all $\overline{\theta}$.*

Proof. See Appendix 7.1. ■

The intuitive explanation behind the above result is the following. Since Cournot competition is by definition softer than Bertrand competition, any rigidity in capacities (or output levels) inherited from autarky is a lesser evil than a constraint on quality choice, as quality is related to consumer tastes *all else equal.* In a market where customers exhibit different preferences, the ability to adjust product characteristics matters more than a flexible scale of production.

Here, we have confined our attention to symmetric technological choices. A fully fledged investigation of asymmetric technological endowments is in Appendix 7.2.

7.4.2 A three-stage game

Now we consider interactions taking place over three stages. In the first, the choice of technology between fixed and variable costs has to be undertaken. Adopting the variable cost technology leaves open the possibility to set either quality or quantity in the second stage, whereas, with the fixed cost technology, the firm is bound to compete in quantities, given the autarchic quality chosen beforehand. In the third stage market competition takes place.

If we go through three-stage games differing for the level of the key parameter $\overline{\theta}$, whose admissible range is $(1, 2]$, we obtain the following:

Proposition 7.3 *i) With homogeneous products and variable cost technology, in most of the admissible range of $\overline{\theta}$, the subgame perfect equilibrium of the three-stage game involves both firms choosing the variable cost technology and the autarchic quality. Then, firms compete in quantities in the market stage with homogeneous products. ii) With* *heterogeneity of products and technologies, we have multiple equilibria with predetermined quality. For high values of $\overline{\theta}$, the three stage game has two subgame perfect equilibria. Firms operate with heterogeneous technologies, set qualities in autarky and then compete à la Cournot with differentiated products. iii) With heterogeneity of products and technologies, we have multiple equilibria with mixed predetermined variables. For $\overline{\theta}$ near to the upper bound of the admissible range, the three stage game has two subgame perfect equilibria. Firms operate with heterogeneous technologies, set different variables in autarky and use different controls in the market game.*

Proof. See Appendix 7.3. ■

7.4.3 Discussion

We are investigating the coordination of equilibria in terms of technologies and quality standards. Moreover, we evaluate the time consistency of the choice of technology. Any asymmetric equilibrium is blurred by time inconsistency, since, in autarky, the variable (fixed) cost technology is chosen when $1 < \overline{\theta} < 16/9 \cong 1.777$ ($\overline{\theta} < 16/9$).

In most cases (i.e. when $1 < \overline{\theta} < 1.498$), endogenous coordination over both technology and quality standard obtains. Then, if the market is not extremely affluent, firms tend to adopt a highly competitive stance, supplying homogeneous goods produced through the most efficient technology. Coordination and time consistency are there.

As consumers get richer ($\theta \in [1.498, 1.919]$), the unique subgame perfect equilibrium disappears. The common standard is lost since more affluence leads firms to diverge in qualities and technologies, adopting thus a less competitive behavior. This, however, is not the whole story. Since countries are equal, both firms should adopt the most efficient technology, i.e., v, as long as $1 < \overline{\theta} < 48/27$. If that is the case, we ought to confine to a two-stage game which corresponds to the south-east quadrant of matrix 7.3 (in Appendix 7.3), where the subgame perfect equilibrium in dominant strategies is $\{(q_A, q_A), (x, x)\}$, where firms produce same quality and quantity. This is the only way to get time consistency in this range of $\overline{\theta}$. The pair of strategies (f, f) is never subgame perfect, no matter how the matrix is reduced.

For even higher levels of the marginal willingness to pay ($\overline{\theta} \in (1.919, 2]$), we still get two subgame perfect equilibria of the three-stage game, where firms fail to coordinate in all stages and thus behave inconsistently. However, different results obtain if we go through two subgames. The first, represented by the north-west quadrant of matrix 7.3, obtains by dropping strategy x_A.

The equilibrium in dominant strategies is $\{(f, f), (x, x)\}$, which is time consistent since it can be interpreted as the legacy of an autarchic choice. The second subgame, represented by the south-east quadrant of matrix 7.3, has an equilibrium in dominant strategies, $\{(q_A, q_A), (x, x)\}$. Firms prefer the latter time inconsistent equilibrium.

7.4.4 Welfare assessment

We now turn to social welfare, confining our attention to those market configurations that are candidates as subgame perfect equilibria of the game between firms. The structure of the game is the same as above, except that the payoffs relevant for governments are social welfare levels. We claim:

Proposition 7.4 *For most values of $\overline{\theta}$, the duopoly game equilibrium coincides with that selected by governments aiming at noncooperatively maximizing social welfare in their respective countries. This equilibrium is associated to common standards both in technology and in quality.*[12] *When $\overline{\theta} \in [1.498, 1.919]$, governments face the same coordination problem as firms. No common standard arises endogenously, either in technology or in quality. When $\overline{\theta}$ is high, there is a unique equilibrium of the governments' game, where a common standard for both technology and quality is adopted.*

Proof. See Appendix 7.4 (*I, II, III*). ■

It seems that, for all levels of $\overline{\theta}$, governments should encourage firms to be flexible in terms of quantities. Market power of firms will be weakened. If that is the case, the governments' game reduces to a two-stage game. This trick is not available in the duopoly game. As a way to implementing such a policy of welfare maximization and/or common standards, one may establish common quality standards in autarky in both countries. This can be the result only of a coordinated policy of governments. Once firms are confined to compete *à la* Cournot in the market stage, they endogenously coordinate their strategies over the equilibrium that would be selected by governments. The adoption of quality standards in open economies turns out to be a device for enhancing competition through standardization.[13] A

[12] In this case international coordination may be supported as a non-cooperative equilibrium of the policy game.

[13] A related literature deals with Minimum Quality Standards (see Ronnen, 1991; and Crampes and Hollander, 1995). The role of MQS in open economies with imperfect competition has been investigated by Boom (1995) and Lutz (2000). The possibility that an MQS policy yields long run pro-competitive effects is dealt with by Ecchia and Lambertini (1997).

consequence of that is the complete absence of time inconsistency in the choice of standards. With free trade, governments would have no incentive to renege the standard adopted in autarky.

7.5 Conclusions

In this chapter we have examined, in a partial equilibrium framework, the effects of different sorts of competition on (i) quality standards; (ii) technology adoption; and (iii) social welfare, when vertically differentiated goods are traded among similar countries.

First, we have considered price competition between firms operating with technologies characterized by different degrees of flexibility and efficiency. When firms are allowed to set the timing of moves in the free trade market game, they select sequential play since profits are larger than with simultaneous play. The social loss caused by Stackelberg competition in prices makes room for contingent trade policies such as antidumping duties. The goal is to force producers to move simultaneously and refrain from undercutting, for instance, by constraining them to announce their prices at the very same date. This provides a theoretical case in favor of coordinated trade policy. Surprisingly enough, consumers will not be hurt.

Under Cournot competition, we have a greater variety of controls in the market stage, which, coupled with the endogenous choice of technology, allows for coordination between firms (or countries) over standards on both technology and product design. In such a setting, firms' behavior leads to different equilibrium outcomes, conditional upon consumers' willingness to pay for quality. Over a wide range of the key parameter, competition intensifies. Firms are induced to adopt the same technology and sell homogeneous products, converging to the equilibrium that a social planner should pick out. The decisions of both the social planner and the firms turn out to be time consistent. Such a coincidence disappears if consumers are richer, since competition softens and firms adopt different standards. That is, one of them is time inconsistent. If firms are bound to set quality in autarky or they enjoy patent protection, they non-cooperatively select the socially preferable equilibria independently of consumers' income/preferences. On this ground, we can thus legitimate the adoption by governments of quality standards regardless of whether they are introduced in an autarchic perspective or taking trade into account. The only provision is that they must be adopted by all countries alike, if they are to be beneficial.

Appendices

Appendix 7.1: Proof of Proposition 7.2

(a) Quality competition under capacity constraints. We still consider two equal countries. Firms operate with variable cost technology, as in (7.1). Inverse demand functions are:

$$p_H = \frac{2\overline{\theta}q_H - q_H x_H - q_L x_L}{2};\ p_L = \frac{q_L(2\overline{\theta} - x_H - x_L)}{2}.$$

Profit functions are:

$$\pi_H = \frac{x_H}{2}(2\overline{\theta}q_H - q_H x_H - q_L x_L - 2tq_H^2);\ \pi_L = \frac{q_L x_L}{2}(2\overline{\theta} - x_H - x_L - 2tq_L).$$

If both quantities are fixed at the autarky level, i.e., $x_i = \overline{\theta}/3$, optimal qualities are $q_H = 5\overline{\theta}/12$ and $q_L = \overline{\theta}/3$, as in autarky, due to absence of strategic interaction. The low-quality good is then sold at the autarchic price (7.4), while $p_H = 7\overline{\theta}^2/24$. Equilibrium profits are $\pi_H^{vv}(x_A) = 17\overline{\theta}^3/432$ and $\pi_L^{vv}(x_A) = \overline{\theta}^3/27$, where superscript vv indicates that both firms operate with a variable cost technology, while x_A means that quantity is fixed under autarky. Consumer surplus is $CS_i = 17\overline{\theta}^3/864$ in both countries. Then, social welfare is higher in country H.
(b) Quantity competition under quality constraints. Both goods are sold at the same price, since qualities coincide: $p_H = p_L = \overline{\theta}(2\overline{\theta} - x_H - x_L)/6$. Optimal quantities are $x_H = x_L = 4\overline{\theta}/9$, while profits are $\pi_H^{vv}(q_A) = \pi_L^{vv}(q_A) = 8\overline{\theta}^3/243$, where q_A means that quality corresponds to the autarky level. Again, consumer surplus is the same in both countries, $CS_i = 8\overline{\theta}^3/243$, as it is for social welfare.
(c) Cournot competition with fixed costs. Profit functions are:

$$\pi_H = \frac{\overline{\theta}^2}{64}(8\overline{\theta}x_H - \overline{\theta}^2 - 4x_H x_L - 4x_H^2);\ \pi_L = \frac{\overline{\theta}^2}{64}(8\overline{\theta}x_L - \overline{\theta}^2 - 4x_H x_L - 4x_L^2).$$

Optimal quantities and profits respectively are $x_H = x_L = 2\overline{\theta}/3$ and $\pi_H^{ff}(q_A) = \pi_L^{ff}(q_A) = 7\overline{\theta}^4/576$. Of course, $p_H = p_L = \overline{\theta}^3/24$, and $CS_i = \overline{\theta}^4/36$.

Appendix 7.2: Asymmetric choices with variable costs

(a) Variable high quality, fixed low quality. We are back to firms with variable cost technology. One firm sets her low quality in autarky, while the high-

quality firm is constrained to the same quantity as in autarky. Hence, $q_L = \overline{\theta}/3 = x_H$. Profit functions are:

$$\pi_H = \frac{\overline{\theta}}{18}(5\overline{\theta}q_H - 6q_H^2 - \overline{\theta}x_L);\ \pi_L = \frac{\overline{\theta}}{6}x_L(\overline{\theta} - x_L).$$

At equilibrium, $q_H = 5\overline{\theta}/12$, $x_L = \overline{\theta}/2$, while profits are

$$\pi_H^{vv}(x_A) = 13\overline{\theta}^3/432;\ \pi_L^{vv}(q_A) = \overline{\theta}^3/24,$$

with $\pi_H^{vv}(x_A) < \pi_L^{vv}(q_A)$. The low-quality firm makes more profits than the high-quality firm. By free-riding over the rival's output constraint it takes advantage of the high-quality firm's inability to expand production as market size increases with trade. Then, firm L's market share increases and consumer surplus is $CS_i = 13\overline{\theta}^3/432$.
(b) High quality fixed, low quality variable. Here is the opposite case. The high-quality (resp., low-quality) firm sets quality (resp., quantity) in autarky, with $q_H = \overline{\theta}/3 = x_L$. Profit functions are:

$$\pi_H = \frac{\overline{\theta}}{18}x_H(4\overline{\theta} - 3q_L - 3x_H);\ \pi_L = \frac{\overline{\theta}}{18}q_L(5\overline{\theta} - 6q_L - 3x_H).$$

Optimal controls are $x_H = 11\overline{\theta}/21$ and $q_L = 2\overline{\theta}/7$. Equilibrium profits are $\pi_H^{vv}(q_A) = 121\overline{\theta}^3/2646$; $\pi_L^{vv}(x_A) = 4\overline{\theta}^3/147$. Again, consumers in both countries enjoy the same surplus, $CS_i = 295\overline{\theta}^3/10584$. Since $\pi_H^{vv}(q_A) > \pi_L^{vv}(x_A)$, social welfare is higher in the country of firm H.
(c) High quality with variable costs, low quality with fixed costs, and Cournot competition. In this mixed case, the high-quality good comes from the variable cost technology. The opposite cannot obtain, because fixed technology is less efficient and, therefore, confined to the low quality. Qualities are set in autarky, and firms optimize over quantities. Profit functions are:

$$\pi_H = \frac{\overline{\theta}x_H}{144}(32\overline{\theta} - 24x_H - 9\overline{\theta}x_L);\ \pi_L = \frac{\overline{\theta}^2}{64}(8\overline{\theta}x_L - \overline{\theta}^2 - 4x_Hx_L - 4x_L^2).$$

Equilibrium quantities and profits are:

$$x_H = \frac{2\overline{\theta}(9\overline{\theta} - 32)}{9\overline{\theta} - 96};\ x_L = \frac{64\overline{\theta}}{96 - 9\overline{\theta}};$$

$$\pi_H^{vf}(q_A) = \frac{2\overline{\theta}^3(9\overline{\theta} - 32)^2}{27(3\overline{\theta} - 32)^2};\ \pi_L^{vf}(q_A) = \frac{\overline{\theta}^4(224 - 9\overline{\theta})(32 + 9\overline{\theta})}{576(3\overline{\theta} - 32)^2}.$$

Consumer surplus, equal in both countries, is $CS_i^{vf}(q_A) = \overline{\theta}^3(1024 + 576\overline{\theta} - 135\overline{\theta}^2)/\left[54(3\overline{\theta} - 32)^2\right]$. From the above expressions, we see that $\pi_L^{vf}(q_A) > \pi_H^{vf}(q_A)$ if $\overline{\theta} > 1.509$. Since the two countries differ only for profits, the inequality extends to respective social welfare. As to quantities, $x_L > x_H$ for all acceptable values of $\overline{\theta}$.

(d) High quality with variable costs and quantity constraint, low quality with fixed costs and quality constraint. The high quality firm with variable cost technology supplies the same quantity as in autarky, while the low quality is set in autarky and produced by fixed cost technology. Profit functions are:

$$\pi_H = \frac{\overline{\theta}(40\overline{\theta}q_H - 48q_H^2 - 3\overline{\theta}^2 x_L)}{144}; \; \pi_L = \frac{\overline{\theta}^2(6x_L - \overline{\theta})(3 - 2x_L)}{192}.$$

At equilibrium $q_H = 5\overline{\theta}/12$, and $x_L = 5\overline{\theta}/6$, whereas profits are $\pi_H^{vf}(x_A) = 5\overline{\theta}^3(10 - 3\overline{\theta})/864$ and $\pi_L^{vf}(q_A) = \overline{\theta}^4/36$. If $1 < \overline{\theta} < 1.282$, then $\pi_H^{vf}(x_A) > \pi_L^{vf}(q_A)$. Consumer surplus is $CS_i^{vf}(x_A, q_A) = 5\overline{\theta}^3(8 + 27\overline{\theta})/6912$.

We sum up the above cases in:

Proposition 7.5 *In the asymmetric setting, the profit ranking is not invariant with respect to $\overline{\theta}$, i.e., there is no dominance in the choice of controls.*

Appendix 7.3: A three stage game

Matrix 7.3 describes the three-stage game in normal form. Firms are labelled as 1 and 2, since their location along the quality spectrum depends upon the specific subgame considered. In each cell, the first payoff refers to firm 1, the second to firm 2.

			2		
			f	v	
			q_A	q_A	x_A
	f	q_A	$\pi_L^{ff}(q_A, q_A); \pi_H^{ff}(q_A, q_A)$	$\pi_L^{vf}(q_A, q_A); \pi_H^{vf}(q_A, q_A)$	$\pi_L^{vf}(q_A, x_A); \pi_H^{vf}(q_A, x_A)$
1	v	q_A	$\pi_H^{vf}(q_A, q_A); \pi_L^{vf}(q_A, q_A)$	$\pi_H^{vv}(q_A, q_A); \pi_L^{vv}(q_A, q_A)$	$\pi_H^{vv}(q_A, x_A); \pi_L^{vv}(q_A, x_A)$
		x_A	$\pi_H^{vf}(x_A, q_A); \pi_L^{vf}(x_A, q_A)$	$\pi_H^{vv}(x_A, q_A); \pi_L^{vv}(x_A, q_A)$	$\pi_H^{vv}(x_A, x_A); \pi_L^{vv}(x_A, x_A)$

Matrix 7.3

I) When $1 < \overline{\theta} < 1.498$, we can order the payoffs in matrix 7.3 according to the following inequalities:

$$\pi_H^{vf}(q_A, q_A) > \pi_H^{vv}(q_A, x_A) > \pi_L^{vv}(x_A, q_A) > \pi_H^{vf}(q_A, x_A) =$$
$$= \pi_H^{vf}(x_A, q_A) > \pi_H^{vv}(x_A, x_A) = \pi_L^{vv}(x_A, x_A) > \pi_H^{vv}(q_A, q_A) =$$
$$= \pi_L^{vv}(q_A, q_A) > \pi_H^{vv}(x_A, q_A) > \pi_L^{vf}(x_A, q_A) = \pi_L^{vf}(q_A, x_A) >$$
$$> \pi_L^{vv}(q_A, x_A) > \pi_L^{vf}(q_A, q_A) > \pi_H^{ff}(q_A, q_A) = \pi_L^{ff}(q_A, q_A).$$

Then, the three-stage game has a unique subgame perfect equilibrium given by the triple pairs of sequentially chosen strategies $\{(v,v), (q_A, q_A),(x,x)\}$. The corresponding outcome is $\pi_H^{vv}(q_A, q_A) = \pi_L^{vv}(q_A, q_A)$. This is a perfectly symmetric equilibrium where firms become indistinguishable. The subgame perfect equilibrium of the three-stage game is also the equilibrium of the subgame where strategy x_A is absent, and then it becomes redundant.

II) When $\overline{\theta} \in [1.498, 1.919]$, we have the following inequalities:

$$\pi_L^{vf}(x_A, q_A) = \pi_L^{vf}(q_A, x_A) > \pi_H^{vv}(q_A, x_A) > \pi_L^{vv}(x_A, q_A) >$$
$$> \pi_H^{vv}(x_A, x_A) > \pi_L^{vf}(q_A, q_A) > \pi_L^{vv}(x_A, x_A) > \pi_H^{vv}(q_A, q_A) =$$
$$= \pi_L^{vv}(q_A, q_A) > \pi_H^{vv}(x_A, q_A) > \pi_H^{vf}(q_A, q_A) > \pi_H^{vf}(x_A, q_A) =$$
$$= \pi_H^{vf}(q_A, x_A) > \pi_L^{vv}(q_A, x_A) > \pi_H^{ff}(q_A, q_A) = \pi_L^{ff}(q_A, q_A).$$

Then, we have two subgame perfect equilibria of the three-stage game, represented by the triple pairs $\{(v, f), (q_A, q_A), (x, x)\}$; $\{(f, v), (q_A, q_A), (x, x)\}$. Again, strategy x_A can be disregarded. The equilibrium belongs to the subgame where firms set qualities in autarky. Firms are unable to coordinate over technological choices and quality standards.

III) When $\overline{\theta} \in (1.919, 2]$, we have the following inequalities:

$$\pi_L^{vf}(x_A, q_A) = \pi_L^{vf}(q_A, x_A) > \pi_L^{vf}(q_A, q_A) > \pi_H^{vv}(q_A, x_A) >$$
$$> \pi_L^{vv}(x_A, q_A) > \pi_H^{vv}(x_A, x_A) > \pi_L^{vv}(x_A, x_A) > \pi_H^{vv}(q_A, q_A) =$$
$$= \pi_L^{vv}(q_A, q_A) > \pi_H^{vv}(x_A, q_A) > \pi_L^{vv}(q_A, x_A) > \pi_H^{vf}(q_A, x_A) =$$
$$= \pi_H^{vf}(x_A, q_A) > \pi_H^{ff}(q_A, q_A) = \pi_L^{ff}(q_A, q_A) > \pi_H^{vf}(q_A, q_A).$$

The three-stage game has two subgame perfect equilibria: firms choose distinct technologies and use different controls as a result of the autarky commitments. The outcome is the triple pairs

$$\{(f, v), (q_A, x_A), (x, q)\}; \ \{(v, f), (x_A, q_A), (q, x)\}.$$

Notice that the equilibrium of the three-stage game does not coincide with the equilibrium of the subgame where firms are constrained to adopt quantity as the control variable in the market stage. This subgame has a unique symmetric equilibrium, $\{(f, f), (q_A, q_A), (x, x)\}$, which is the result of the adoption of dominant strategies in a prisoner's dilemma situation.

Appendix 7.4: Welfare comparisons

Social welfare levels are in matrix 7.4. The equilibria of the game can be singled out from the sequence of inequalities, specified for the relevant intervals of $\overline{\theta}$.

			2		
			f	v	
			q_A	q_A	x_A
1	f	q_A	$W_1^{ff}(q,q); W_2^{ff}(q,q)$	$W_1^{vf}(q,q); W_2^{vf}(q,q)$	$W_1^{vf}(q,x); W_2^{vf}(q,x)$
	v	q_A	$W_1^{vf}(q,q); W_2^{vf}(q,q)$	$W_1^{vv}(q,q); W_2^{vv}(q,q)$	$W_1^{vv}(q,x); W_2^{vv}(q,x)$
		x_A	$W_1^{vf}(x,q); W_2^{vf}(x,q)$	$W_1^{vv}(x,q); W_2^{vv}(x,q)$	$W_1^{vv}(x,x); W_2^{vv}(x,x)$

Matrix 7.4

I) $1 < \overline{\theta} < 1.498$. In this range of the marginal willingness to pay, social welfare ranks as follows:

$$W_1^{vf}(q,q) > W_1^{vv}(q,x) > W_1^{vv}(q,q) > W_1^{vf}(x,q) > W_1^{vv}(x,q) > W_1^{vv}(x,x)$$
$$> W_1^{vf}(q,x) > W_1^{vf}(q,q) > W_1^{ff}(q,q).$$
$$W_2^{vf}(q,q) > W_2^{vv}(x,q) > W_2^{vv}(q,q) > W_2^{vf}(q,x) > W_2^{vv}(x,x) > W_2^{vv}(q,x)$$
$$> W_2^{vf}(x,q) > W_2^{vf}(q,q) > W_2^{ff}(q,q).$$

The subgame perfect equilibrium, $\{(v, v), (q_A, q_A), (x, x)\}$, coincides with the equilibrium of the game between firms. Moreover, this is also the equilibrium of the subgame which obtains by deleting strategy x_A.

II) $\overline{\theta} \in [1.498, 1.919]$. In this range, the relevant inequalities are:

$$W_1^{vf}(q,x) > W_1^{vf}(q,q) > W_1^{vv}(q,x) > W_1^{vf}(q,q) > W_1^{vf}(x,q) > W_1^{ff}(q,q)$$

$$> W_1^{vv}(q,q) > W_1^{vv}(x,q) > W_1^{vv}(x,x).$$

$$W_2^{vf}(x,q) > W_2^{vf}(q,q) > W_2^{vv}(x,q) > W_2^{vf}(q,q) > W_2^{vf}(q,x) > W_2^{ff}(q,q)$$
$$> W_2^{vv}(q,q) > W_2^{vv}(x,x) > W_2^{vv}(q,x).$$

This game has two subgame perfect equilibria,

$$\{(f,v),(q_A,q_A),(x,x)\} \text{ and } \{(v,f),(q_A,q_A),(x,x)\},$$

coinciding with those of the duopoly market game. In each equilibrium, social welfare of the country operating with the fixed cost technology is higher than the other country's.

III) $\overline{\theta} \in]1.919, 2]$. When the marginal willingness to pay for quality is high, we have:

$$W_1^{vf}(q,x) > W_1^{vf}(q,q) > W_1^{ff}(q,q) > W_1^{vv}(q,x) > W_1^{vf}(x,q) > W_1^{vf}(q,q)$$
$$> W_1^{vv}(q,q) > W_1^{vv}(x,q) > W_1^{vv}(x,x).$$
$$W_2^{vf}(x,q) > W_2^{vf}(q,q) > W_2^{ff}(q,q) > W_2^{vv}(x,q) > W_2^{vf}(q,x) > W_2^{vf}(q,q)$$
$$> W_2^{vv}(q,q) > W_2^{vv}(x,x) > W_2^{vv}(q,x).$$

The game yields $\{(f,f),(q_A,q_A),(x,x)\}$ as its unique equilibrium in dominant strategies. The fixed cost technology is adopted and quality is set in autarky in both countries. There is no coincidence between the duopoly game and the governments' game equilibria.

References

1] Boom, A. (1995), "Asymmetric International Minimum Quality Standards and Vertical Differentiation", *Journal of Industrial Economics*, **43**, 101-19.
2] Bronckers, M. (1996), "Rehabilitating Antidumping and Other Trade Remedies through Cost-Benefit Analysis", *Journal of World Trade. Law, Economics, Public Policy*, **30**, 6-37.
3] Crampes, C. and A. Hollander (1995), "Duopoly and Quality Standards", *European Economic Review*, **39**, 71-82.
4] Davidson, C. and R. Deneckere (1986), "Long-run Competition in Capacity, Short-run Competition in Price, and the Cournot Model", *RAND Journal of Economics*, **17**, 404-15.

5] Ecchia, G. and L. Lambertini (1997), "Minimum Quality Standards and Collusion", *Journal of Industrial Economics*, **45**, 101-13.
6] Farrell, J. and G. Saloner (1985), "Standardization, Compatibility, and Innovation", *RAND Journal of Economics*, **16**, 70-83.
7] Farrell, J. and G. Saloner (1987), "The Economics of Horses, Penguins, and Lemmings", in L.G. Gable (ed.), *Production Standardization and Competitive Strategies*, Amsterdam, North-Holland.
8] Hamilton, J. and S. Slutsky (1990), "Endogenous Timing in Duopoly Games: Stackelberg or Cournot Equilibria", *Games and Economic Behavior*, **2**, 29-46.
9] Jensen, R. and M. Thursby (1996), "Patent Races, Product Standards, and International Competition", *International Economic Review*, **37**, 21-49.
10] Katz, M. and C. Shapiro (1985), "Network Externalities, Competition, and Compatibility", *American Economic Review*, **75**, 424-40.
11] Katz, M. and C. Shapiro (1986), "Technology Adoption in the Presence of Network Externalities", *Journal of Political Economy*, **94**, 822-41.
12] Kreps, D. and J. Scheinkman (1983), "Quantity Precommitment and Bertrand Competition Yield Cournot Outcomes", *Bell Journal of Economics*, **14**, 326-37.
13] Lambertini, L. (1997), "Intraindustry Trade under Vertical Product Differentiation", *Keio Economic Studies*, **34**, 51-69.
14] Lutz, S. (2000), "Trade Effects of Minimum Quality Standards with and without Deterred Entry", *Journal of Economic Integration*, **15**, 314-44.
15] Matutes, C. and P. Regibeau (1988), "Mix and Match: Product Compatibility Without Network Externalities", *RAND Journal of Economics*, **19**, 221-34.
16] Motta, M. (1992), "Sunk Costs and Trade Liberalization", *Economic Journal*, **102**, 578-87.
17] Motta, M., J.-F. Thisse and A. Cabrales (1997), "On the Persistence of Leadership or Leapfrogging in International Trade", *International Economic Review*, **38**, 809-24.
18] Mussa, M. and S. Rosen (1978), "Monopoly and Product Quality", *Journal of Economic Theory*, **18**, 301-17.
19] Ronnen, U. (1991), "Minimum Quality Standards, Fixed Costs, and Competition", *RAND Journal of Economics*, **22**, 490-504.
20] Shaked, A. and J. Sutton (1982), "Relaxing Price Competition through Product Differentiation", *Review of Economic Studies*, **49**, 3-13.
21] Shaked, A. and J. Sutton (1983), "Natural Oligopolies", *Econometrica*, **51**, 1469-83.
22] Shaked, A. and J. Sutton (1984), "Natural Oligopolies and International Trade", in H. Kierzkowski (ed.), *Monopolistic Competition and International*

Trade, Oxford, Clarendon Press.
23] Spence, A.M. (1975), "Monopoly, Quality, and Regulation", *Bell Journal of Economics*, **6**, 417-29.
24] Tirole, J. (1988), *The Theory of Industrial Organization*, Cambridge, MA, MIT Press.
25] Vandenbussche, H. (1995), "How Can Japanese and Central-European Exporters to the EU Avoid Antidumping Duties?", *World Competition, Law and Economics Review*, **18**, 55-73.
26] Vandenbussche, H. (1996), "Is European Antidumping Protection against Central Europe Too High?", *Weltwirtschaftliches Archiv*, **132**, 116-38.
27] Veugelers, R. and Vandenbussche, H. (1999), "European Antidumping Policy and the Profitability of National and International Collusion", *European Economic Review*, **43**, 1-28.

Part II

Dynamics

Chapter 8

RJVs for product innovation and cartel stability

Cristina Iori and Luca Lambertini[1]

8.1 Introduction

Oligopoly theory has produced a relevant literature on repeated market interaction. The relative efficiency of Bertrand and Cournot competition in stabilizing cartels composed by firms whose products are imperfect substitutes has been analysed by Deneckere (1983), Rothschild (1992) and Albæk and Lambertini (1998), showing that when substitutability between products is high, collusion is better supported in price-setting games than in quantity-setting games, while the reverse is true in case of low substitutability.[2] Majerus (1988) has proved that this result is not confirmed as the number of firms increases. These contributions compare Cournot and Bertrand supergames to conclude that a quantity-setting cartel should almost always be preferred to a price-setting cartel on stability grounds.[3] Finally, the influence of endogenous product differentiation on the stability of collusion in prices has

[1]We thank Paolo Garella and Raimondello Orsini for useful comments and discussion. The usual disclaimer applies.

[2]The same question is addressed in Lambertini (1996), where the evaluation of cartel stability under Bertrand and Cournot behaviour is carried out in terms of the concavity/convexity of the market demand function.

[3]This approach cannot grasp any strategic interaction behind the choice of the market variable. Using the same demand structure as in Deneckere (1983) and analysing asymmetric cartels where one firm is a Bertrand agent while the other is a Cournot one, Lambertini (1997) proves that the choice of the market variable in order to stabilize implicit collusion produces a Prisoner's Dilemma.

been investigated by Chang (1991, 1992), Ross (1992) and Häckner (1994, 1995, 1996). The main finding reached by these contributions is that, under vertical differentiation, collusion is more easily sustained, the more similar the products are, while the opposite applies under horizontal differentiation.

The consequences of collusion on the extent of optimal differentiation in the horizontal differentiation model have also received attention. Friedman and Thisse (1993) have considered a repeated price game in the horizontal framework and found out that minimum differentiation obtains if firms collude in the market stage. In most of these models, although differentiation can be endogenously determined by firms through strategic interaction, the issue of cartel stability is studied by making the degree of differentiation vary symmetrically around the ideal midpoint of the interval of technologically feasible or socially preferred varieties, leading to the conclusion that producers may prefer to choose the characteristics of their respective goods differently from what profit maximization would suggest, if this helps them minimize the incentive to deviate from the implicit cartel agreement.

To our knowledge, little attention has been paid so far to the interplay between firms' technological decisions and their ability to build up and maintain collusive agreements over time (one exception being Lambertini, 2000). This is a relevant issue, in that public authorities prosecute collusive market behaviour, while they seldom discourage cooperation in R&D activities. Indeed, there exist many examples of policy measures designed so as to stimulate the formation of research joint ventures.[4] However, encouraging cooperative R&D and discouraging market collusion can be mutually inconsistent moves, if R&D cooperation tends to facilitate collusion in the product market.

In this respect, Martin (1995) analyses the strategic effects of a research joint venture (*RJV* henceforth) designed to achieve a *process* innovation for an existing product. Then, the product is marketed by firms engaging in repeated Cournot behaviour over an infinite time horizon. Martin shows that cooperation in process innovation enhances implicit collusion, which can jeopardise the welfare advantage of eliminating effort duplication through the *RJV*. This result has potential implications for the case of product innovation as well. Lambertini, Poddar and Sasaki (1998, 2003) adopt the same view as Martin (1995), although they consider the relationship between the organization of R&D for product innovation and the stability of implicit cartel agreeements in a duopoly model with linear demand functions *à la*

[4]See the National Cooperative Research Act in the US; EC Commission (1990); and, for Japan, Goto and Wakasugi (1988).

Singh and Vives (1984), finding mixed results.[5]

We reassess Martin's framework, by generalising the analysis presented in Lambertini, Poddar and Sasaki (2002, pp. 840-843). A vertically differentiated market is considered, where firms are given the possibility of choosing between activating either independent ventures leading to distinct product qualities, or a joint venture for a single quality, aimed at reducing the initial R&D expenditure *vis à vis* independent ventures. Then, firms market the product(s) over an infinite horizon. In doing so, they either repeat the one-shot Nash equilibrium forever, or behave collusively, according to their intertemporal discounting. In such a setting, we prove that Martin's conclusion is confirmed, i.e., the RJV makes it easier for firms to sustain collusive behaviour in the market supergame, as compared to independent ventures. This holds independently of whether firms set prices or quantities during the supergame. Our result entails that public policies towards the R&D behaviour of firms should be tailored case by case, so as not to become inconsistent with the pro-competitive attitude characterising the current legislation on marketing practices.

The remainder of the paper is structured as follows. The basic model of vertical differentiation is described in section 8.2. Section 8.3 describes the case of collusion along the frontier of monopoly profits. Sections 8.4 and 8.5 deals with the construction of discounted profit flows under either Cournot or Bertrand behaviour. The equilibrium venture decisions are characterised in section 8.6. Finally, section 8.7 provides concluding remarks.

8.2 The vertical differentiation model

We adopt a well known model of duopoly under vertical differentiation (see Gabszewicz and Thisse, 1979, 1980; Choi and Shin, 1992; Motta, 1993; Aoki and Prusa, 1997; Lehmann-Grube, 1997; Lambertini, 1999, *inter alia*).[6] Two single-product firms, labelled as H and L, produce goods of (different) qualities q_H and $q_L \in [0, \infty)$, with $q_H \geq q_L$, through the same technology, $C(q_i) = cq_i^2$, with $c > 0$. This can be interpreted as fixed cost due to the R&D effort needed to produce a certain quality, while variable production costs are assumed away.[7] Products are offered on a market where consumers

[5] See also Lambertini, Poddar and Sasaki (2002) for an application to the Hotelling duopoly model. Cabral (1996, 2000), in a somewhat dissimilar vein, proves the possibility that competitive pricing is needed to sustain more efficient R&D agreements.

[6] A different model is used in Shaked and Sutton (1982, 1983), where fixed costs are exogenous.

[7] In Lambertini, Poddar and Sasaki (2002), development costs k_i are exogenously given for both firms, with $k_H > k_L$. This assumption, while allowing for a simple parametric

have unit demands, and buy if and only if the net surplus from consumption $v_\theta(q_i, p_i) = \theta q_i - p_i \geq 0$, where p_i is the unit price of the good of quality q_i, purchased by a generic consumer whose marginal willingness to pay is $\theta \in [0, \overline{\theta}]$. We assume that θ is uniformly distributed with density one over such interval, so that the total mass of consumer is $\overline{\theta}$.

Firms interact over $t \in [0, \infty)$, as follows:

- At $t = 0$, they conduct R&D towards the development of product quality, through either a joint venture (RJV) or independent ventures (IV henceforth). If firms undertake a joint venture, then $q_i = q_j = q$ and each firm bears half the development cost, $cq^2/2$. Otherwise, firms market differentiated products, each of them bearing the full development cost of their respective varieties, cq_i^2.[8]

- Over $t \in [1, \infty)$, firms market the product(s) resulting from previous R&D activity, either *à la* Cournot or *à la* Bertrand.

- In the infinitely long marketing phase, firms may collude if their respective time discounting allows them to do so. Otherwise, they always play *à la* Nash. Define as δ_i the discount factor of firm i, and $\delta_i^I(K)$ the critical threshold for the stability of collusion, with superscript $I = B, C$ standing for Bertrand and Cournot, and $K = IV, RJV$, indicating the organizational design chosen for the R&D phase.

As a first step, observe that the locations of indifferent consumers along $[0, \overline{\theta}]$ are:

$$\theta_H = \frac{p_H - p_L}{q_H - q_L} \; ; \; \theta_L = \frac{p_L}{q_L} \tag{8.1}$$

where θ_H is the marginal willingness to pay of the consumer who is indifferent between q_H and q_L; and θ_L is the marginal willingness to pay of the consumer who is indifferent between q_L and not buying at all. Then, market demands are

$$x_H = \overline{\theta} - \theta_H \; ; \; x_L = \theta_H - \theta_L \, . \tag{8.2}$$

analysis of venture decisions, excludes the possibility of investigating the incentives to endogenously adjust the R&D efforts and the resulting quality levels depending upon venture decisions and the price or quantity behaviour.

[8] The R&D efforts of firms operating in vertically differentiated markets are investigated in Beath, Katsoulacos and Ulph (1987), Motta (1992), Dutta, Lach and Rustichini (1995), Rosenkranz (1995, 1997), van Dijk (1996), Lambertini, Poddar and Sasaki (2002). In particular, Motta (1992) and Rosenkranz (1997) describe the incentives towards cooperative R&D.

Notice that (8.2) can be inverted to yield the relevant demand functions for the Cournot case:

$$p_H = q_H \left(\overline{\theta} - x_H\right) - q_L x_L \ ; \ p_L = q_L \left(\overline{\theta} - x_H - x_L\right) \ . \tag{8.3}$$

At any $t \geq 1$, firm i obtains revenues $R_i^I = p_i x_i$, $I = B, C$. The discounted flow of profits over the whole game is then:

$$\pi_i^I = \begin{cases} \dfrac{\delta_i}{1-\delta_i} \cdot R_i^I(q_i, q_j) - cq_i^2 \text{ under } IV \\ \dfrac{\delta_i}{1-\delta_i} \cdot R_i^I(q) - \dfrac{cq^2}{2} \text{ under } RJV \end{cases} \tag{8.4}$$

To model collusion in marketing, we adopt the Perfect Folk Theorem (PFT henceforth; see Friedman, 1971), where the infinite reversion to the one-shot Nash equilibrium is used as a punishment following any deviation from the prescribed collusive path.[9] The collusive path can instruct firms to collude either fully (i.e., at the Pareto frontier of monopoly profits) or partially, at any pair of prices or quantities such that per-period individual revenues are at least as large as the Nash equilibrium revenues.

Define:

[1] The instantaneous best reply of firm i as α_i^*.

[2] The collusive action as $\alpha^{coll} \in \left(\min\left\{\alpha^N, \alpha^M\right\}, \max\left\{\alpha^N, \alpha^M\right\}\right]$, with $\alpha = p, x$.

[3] The collusive revenues to firm i as $R_i^{Icart}(\cdot)$, $(\cdot) = \{(q), (q_i, q_j)\}$.

[4] The one-shot Nash revenues to firm i as $R_i^{IN}(\cdot)$.

[5] The one-shot deviation revenues to firm i as $R_i^{ID}(\cdot)$.

The rules of the PFT establish what follows:

- At $t = 0$, firms play α^{cart}.
- At $t \geq 1$, firms play α^{cart} iff $\alpha_i = \alpha^{cart}$ at $t-1$ for all i ; firms play α_i^* otherwise.

[9] There exist other (less grim) penal codes (see Abreu, 1986; 1988; Abreu, Pearce and Stacchetti, 1986; Fudenberg and Maskin, 1986), using symmetric optimal punishments. However, the asymmetry of our model prevents us from adopting optimal punishments. For the application of optimal punishments in a symmetric duopoly model with product differentiation, see Lambertini and Sasaki (1999, 2002).

Definitions [3-5] and the rules of PFT yields that implicit collusion at α^{coll} is sustainable iff

$$\delta_i \geq \delta_i^I(K) = \frac{R_i^{ID}(\cdot) - R_i^{Icart}(\cdot)}{R_i^{ID}(\cdot) - R_i^{IN}(\cdot)} \quad \text{for all } i \, . \tag{8.5}$$

In the next section, we quickly deal with the case of full collusion, where $\alpha^{cart} = \alpha^M$.

8.3 Stability analysis of collusion in prices or quantities

First, notice that when firms operate along the frontier of monopoly profits, they are indifferent between setting prices or output levels. Therefore, we confine our attention to the Bertrand case.

Suppose firms choose independent ventures at $t = 0$. Then, over $t \in [1, \infty)$, they should market differentiated products. We are going to show that this cannot be an equilibrium. At any $t \in [1, \infty)$, the cartel aims at

$$\max_{p_H, p_L} R^M = R_H^B(q_H, q_L) + R_L^B(q_L, q_H) \, . \tag{8.6}$$

Monopoly prices are:

$$p_H^M = \frac{\overline{\theta} q_H}{2} \, ; \, p_L^M = \frac{\overline{\theta} q_L}{2} \, , \tag{8.7}$$

at which $x_H^M = \frac{\overline{\theta}}{2}$, while $x_L^M = 0$. Therefore,

$$\pi_H^B = \frac{\delta_H}{1 - \delta_H} \cdot \frac{\overline{\theta}^2 q_H}{4} - c q_H^2 \, ; \, \pi_L^B = -c q_L^2 \, . \tag{8.8}$$

On the basis of the above result, independent ventures imply that, for all $q_L \in (0, q_H)$, the low-quality firm would exit, getting thus zero profits. Alternatively, firm L may produce $q_L = q_H$. This immediately entails that $\delta_i^B = 1/2$ for all i, as firms offer homogeneous goods.

It needs no proof to show that the same holds in the case of a joint venture, as this would yield product homogeneity as a result of technological decisions taken at $t = 0$. We have thus proved the following:

Lemma 8.1 *Under full collusion, the low-quality product enjoys zero demand. As a consequence, firms will only supply homogeneous goods.*

The above lemma also implies the following relevant corollary:

Corollary 8.1 *Under full collusion in prices, $\delta_i^B = 1/2$ for all i, independently of firms' venture decisions.*

As to the Cournot case, it can be quickly verified that, for all $q_L \in [0\,,\,q_H)$, we have

$$x_H^M = \frac{\overline{\theta}}{2}\,,\; x_L^M = 0 \tag{8.9}$$

which again entails that the low-quality firm survives only if $q_L = q_H$, either because firms activate an RJV, or because firms develop the same quality independently of each other. As a result, we can state the following:

Lemma 8.2 *Under full collusion in quantities, $\delta_i^C = 9/17$ for all i, independently of firms' venture decisions.*

In summary, independently of the market variable chosen for the supergame over $t \in [1, \infty)$, the firms' venture decisions at $t = 0$ have no bearings on the stability of collusion, as setting either monopoly prices or quantities induces firms to play a supergame with homogeneous goods. Given that the critical threshold is unique in each setting, in the remainder we adopt the simplifying assumption $\delta_H = \delta_L = \delta$.

Having characterised the critical threshold of the discount factors ensuring the stability of collusion in both settings, we can now move on to the construction of the relevant payoffs, i.e., net discounted profit flows, for each of the cases under examination.

8.4 Case I: Collusion

Here, firms' time preferences are such that $\delta \geq \delta^I$, $I =, C$, so that firms collude with homogeneous goods under both IV and RJV. Irrespective of whether firms set prices or quantities, the individual objective function to be maximised w.r.t. quality is either:

$$\pi = \frac{\delta}{1-\delta} \cdot \frac{\overline{\theta}^2 q}{8} - cq^2 \text{ under } IV; \tag{8.10}$$

or

$$\pi = \frac{\delta}{1-\delta} \cdot \frac{\overline{\theta}^2 q}{8} - \frac{cq^2}{2} \text{ under } RJV. \tag{8.11}$$

With independent ventures, the first order condition (FOC) w.r.t. q is:

$$\frac{\partial \pi}{\partial q} = \frac{\delta}{1-\delta} \cdot \frac{\overline{\theta}^2}{8} - 2cq = 0 \tag{8.12}$$

yielding the following equilibrium quality:

$$q^* = \frac{\delta \overline{\theta}^2}{16c(1-\delta)} \tag{8.13}$$

and the corresponding profits:

$$\pi^{cart}(IV) = \frac{\delta^2 \overline{\theta}^4}{256c(1-\delta)^2}. \tag{8.14}$$

If instead firms activate a joint venture, the FOC w.r.t. q is:

$$\frac{\partial \pi}{\partial q} = \frac{\delta}{1-\delta} \cdot \frac{\overline{\theta}^2}{8} - cq = 0 \tag{8.15}$$

yielding:

$$q^* = \frac{\delta \overline{\theta}^2}{8c(1-\delta)}; \ \pi^{cart}(RJV) = \frac{\delta^2 \overline{\theta}^4}{128c(1-\delta)^2}. \tag{8.16}$$

This discussion immediately entails:

Lemma 8.3 *If $\delta \geq \delta^I$, $I =, C$, then $RJV \succ IV$ due to the cost-saving effect.*

8.5 Case II: Competition

In this section, we investigate the situation where $\delta < \delta^I$, $I =, C$, so that firms compete, either in quantities or in prices. Unlike the collusive case, here firms may supply either differentiated or homogeneous goods, depending on the technological decision taken at $t = 0$. First, we characterise the Cournot setting.

8.5.1 Cournot competition

First of all, observe that the market stage can be solved as if firms were playing a one-stage game, since quantities only appear in R_i^C, $i = H, L$. The relevant FOCs are:

$$\frac{\partial \pi_H^C}{\partial x_H} = \frac{\delta}{1-\delta} \cdot \frac{\partial R_H^C}{\partial x_H} = \frac{\delta}{1-\delta} \left[q_H \left(\overline{\theta} - 2x_H \right) - q_L x_L \right] = 0\,; \quad (8.17)$$

$$\frac{\partial \pi_L^C}{\partial x_L} = \frac{\delta}{1-\delta} \cdot \frac{\partial R_L^C}{\partial x_L} = \frac{\delta}{1-\delta} q_L \left(\overline{\theta} - x_H - 2x_L \right) = 0\,. \quad (8.18)$$

The system (8.17-8.18) can be solved to obtain optimal outputs for a generic quality pair (cf. Motta, 1993):

$$x_H^N = \frac{\overline{\theta}\,(2q_H - q_L)}{4q_H - q_L}\,;\; x_L^N = \frac{\overline{\theta} q_H}{4q_H - q_L}\,. \quad (8.19)$$

Independent ventures

Here, each firm noncooperatively develops her own product quality, $q_H \neq q_L$. Using (8.19), we can simplify the expressions of the instantaneous revenues:

$$R_H^C = \frac{\overline{\theta}^2 q_H \left(2q_H - q_L\right)^2}{\left(4q_H - q_L\right)^2}\,;\; R_L^C = \frac{\overline{\theta}^2 q_H^2 q_L}{\left(4q_H - q_L\right)^2}\,. \quad (8.20)$$

In order to characterise the optimal quality choices of the two firms at $t = 0$, it is useful to adopt the following transformation of variables. Define:

$$\Psi_i^C \equiv \frac{(1-\delta)}{\delta} \pi_i^C \quad (8.21)$$

so that one can write:

$$\Psi_i^C = R_i^C - \gamma q_i^2\,, \text{ where } \gamma \equiv \frac{c\,(1-\delta)}{\delta}\,. \quad (8.22)$$

This makes it possible to proceed along the same line as in Motta (1993), to solve the quality stage. The relevant FOCs are:

$$\frac{\partial \Psi_H^C}{\partial q_H} = \frac{\overline{\theta}^2 \left[(2q_H - q_L)\,(8q_H^2 - 2q_H q_L + q_L^2) - 2\gamma\,(4q_H - q_L)^3 q_H \right]}{(4q_H - q_L)^3} = 0\,; \quad (8.23)$$

$$\frac{\partial \Psi_L^C}{\partial q_L} = \frac{\overline{\theta}^2 q_H^2\,(4q_H + q_L)^3 - 2\gamma\,(4q_H - q_L)^3 q_L}{(4q_H - q_L)^3} = 0\,, \quad (8.24)$$

yielding:

$$q_H^{CN} = 0.12597 \frac{\overline{\theta}^2}{\gamma} = 0.12597 \frac{\delta \overline{\theta}^2}{c\,(1-\delta)}\,; \quad (8.25)$$

$$q_L^{CN} = 0.04511 \frac{\overline{\theta}^2}{\gamma} = 0.04511 \frac{\delta \overline{\theta}^2}{c\,(1-\delta)}\,. \quad (8.26)$$

The associated equilibrium profits are:

$$\pi_H^{CN}(IV) = 0.00974\frac{\delta^2\overline{\theta}^4}{c(1-\delta)^2}; \quad (8.27)$$

$$\pi_L^{CN}(IV) = 0.00137\frac{\delta^2\overline{\theta}^4}{c(1-\delta)^2}. \quad (8.28)$$

Joint venture

Here, we can impose $q_H = q_L = q$. Then, equilibrium quantities can be rewritten as follows:

$$x_H^N = x_L^N = \frac{\overline{\theta}}{3}. \quad (8.29)$$

Consequently, the maximand of firm i at $t = 0$ is:

$$\pi = \frac{\delta}{1-\delta}\cdot\frac{\overline{\theta}^2 q}{9} - \frac{cq^2}{2}, \quad (8.30)$$

where we have dropped the index i as it is irrelevant. The FOC w.r.t q is:

$$\frac{\partial\pi}{\partial q} = \frac{\delta}{1-\delta}\cdot\frac{\overline{\theta}^2}{9} - cq = 0 \quad (8.31)$$

yielding the optimal quality $q^*(RJV) = \delta\overline{\theta}^2 / [9(1-\delta)]$. The individual equilibrium profits are:

$$\pi^{CN}(RJV) = \frac{\delta^2\overline{\theta}^4}{162c(1-\delta)^2}. \quad (8.32)$$

8.5.2 Bertrand competition

The case where firms set prices can be treated much the same way as the Cournot setting. FOCs w.r.t. prices are:

$$\frac{\partial\pi_H^C}{\partial p_H} = \frac{\delta}{1-\delta}\cdot\frac{\partial R_H^C}{\partial p_H} = \frac{\delta}{1-\delta}\cdot\frac{\left[p_L - 2p_H + \overline{\theta}(q_H - q_L)\right]}{(q_H - q_L)} = 0; \quad (8.33)$$

$$\frac{\partial\pi_L^C}{\partial p_L} = \frac{\delta}{1-\delta}\cdot\frac{\partial R_L^C}{\partial p_L} = \frac{\delta}{1-\delta}\cdot\frac{(2p_L q_H - p_H q_L)}{q_L(q_H - q_L)} = 0. \quad (8.34)$$

Solving the system (8.33-8.34) yields the Nash equilibrium prices for any given quality pair (again, cf. Motta, 1993):

$$p_H^N = \frac{2\overline{\theta}q_H(q_H - q_L)}{4q_H - q_L}; \; p_L^N = \frac{\overline{\theta}q_L(q_H - q_L)}{4q_H - q_L}. \quad (8.35)$$

Independent ventures

Assume $q_H \neq q_L$. Then, using the transformation:

$$\gamma \equiv \frac{c(1-\delta)}{\delta} \Leftrightarrow \Psi_i^B \equiv \frac{(1-\delta)}{\delta}\pi_i^B \tag{8.36}$$

we can write the FOCs pertaining to the quality stage at $t = 0$, as follows:

$$\frac{\partial \Psi_H^B}{\partial q_H} = \frac{4\overline{\theta}^2 q_H (4q_H^2 - 3q_H q_L + 2q_L^2) - 2\gamma (4q_H - q_L)^3 q_H}{(4q_H - q_L)^3} = 0; \tag{8.37}$$

$$\frac{\partial \Psi_L^B}{\partial q_L} = \frac{\overline{\theta}^2 q_H^2 (4q_H - 7q_L) - 2\gamma (4q_H - q_L)^3 q_L}{(4q_H - q_L)^3} = 0, \tag{8.38}$$

yielding:

$$q_H^{BN} = 0.12665\frac{\overline{\theta}^2}{\gamma} = 0.12665\frac{\delta\overline{\theta}^2}{c(1-\delta)}; \tag{8.39}$$

$$q_L^{BN} = 0.02411\frac{\overline{\theta}^2}{\gamma} = 0.02411\frac{\delta\overline{\theta}^2}{c(1-\delta)}. \tag{8.40}$$

As a result, equilibrium profits are:

$$\pi_H^{BN} = 0.01222\frac{\delta^2\overline{\theta}^4}{c(1-\delta)^2}; \tag{8.41}$$

$$\pi_L^{BN} = 0.00076\frac{\delta^2\overline{\theta}^4}{c(1-\delta)^2}. \tag{8.42}$$

Joint venture

In this case, things are a little bit more involved than in the previous ones. The reason is that, firms being unable to collude, the resulting revenues are nil in every period as a consequence of the Bertrand-Nash equilibrium in prices associated with homogeneous goods, which, in turn, is the direct implication of the joint venture. This, of course, implies that, if firms decide to stay in the market, then the resulting quality tends to zero in order to reduce as much as possible the loss caused by the fixed R&D cost. Alternatively, both firms may exit, or just one of them does, leaving full monopoly power to the other one. In such a case, the only incumbent (identified by superscript M) reoptimises its price and quality levels so as to maximise:

$$\pi^M = \frac{\delta}{1-\delta}R^M - cq^2, \tag{8.43}$$

where:

$$R^M = px\,;\; x = \overline{\theta} - \frac{p}{q}\,. \tag{8.44}$$

The relevant FOCs are:

$$\frac{\partial \pi^M}{\partial p} = \frac{\delta}{1-\delta} \cdot \frac{\partial R^M}{\partial p} = \frac{\delta}{1-\delta} \cdot \frac{\left(\overline{\theta} q - 2p\right)}{q} = 0 \tag{8.45}$$

yielding $p^M = \overline{\theta} q/2$; and:

$$\frac{\partial \pi^M}{\partial q} = \frac{\delta \overline{\theta}^2 - 8c\,(1-\delta)\,q}{4\,(1-\delta)} = 0 \tag{8.46}$$

yielding $q^M = \delta \overline{\theta}^2 / \left[8c\,(1-\delta)\right]$. The associated monopoly profits are $\pi^M = \delta^2 \overline{\theta}^4 / \left[64c\,(1-\delta)^2\right]$.

As to the stay-exit decision, it may be summarised by the 'exit game' in matrix 8.1 below, where s stands for *stay* while e stands for *exit*.

		firm L	
		s	e
firm H	s	$\sim 0\,;\sim 0$	$\pi^M\,;\,0$
	e	$0\,;\,\pi^M$	$0\,;\,0$

Matrix 8.1: The exit game

In the north-west cell, the payoff ~ 0 indicates that profits are approximately nil as the joint venture must reduce to a minimum the loss due to R&D costs. This game has obviously two pure-strategy Nash equilibria, namely, (s, e) and (e, s). Moreover, there also exists a mixed-strategy equilibrium, that can be quickly calculated as follows. Suppose firm L plays s with probability $\mathfrak{p}$ and e with probability $1 - \mathfrak{p}$. Then, the expected profits accruing to firm H are, respectively:

$$E\left(\pi\left(s\right)\right) = \left(1 - \mathfrak{p}\right)\pi^M\,;\; E\left(\pi\left(e\right)\right) = \mathfrak{p}\pi^M\,. \tag{8.47}$$

Accordingly, by imposing $E\left(\pi\left(s\right)\right) = E\left(\pi\left(e\right)\right)$ and solving for $\mathfrak{p}$ one finds the probability that firm L must associate with playing strategy s so as to make firm H indifferent between staying and exiting. In view of the symmetry of the model, such a probability is of course $\mathfrak{p} = 1/2$. The same considerations obviously hold for the probability of firm H staying in the market.

As a result of the foregoing discussion, we can conclude that the relevant (expected) individual profits in the case of a Bertrand game following a joint venture are:

$$\pi_H^{BN}(RJV) = \pi_L^{BN}(RJV) = \pi^{BN}(RJV) = \frac{\pi^M}{4} = \frac{\delta^2\overline{\theta}^4}{256c(1-\delta)^2}. \quad (8.48)$$

8.6 Optimal venture decisions

Now we are in a position to solve the venture stage at $t = 0$. As we have explained at the outset, this decision is taken by firms non-cooperatively and simultaneously at the first stage of the investment game taking place at the initial instant. We analyse first the technological commitments taken by firms when the investment in quality is followed by a Cournot supergame.

8.6.1 Venture commitments in the Cournot setting

Here, we know that firms will (i) compete for all $\delta \in [0, 9/17)$; (ii) collude for all $\delta \in [9/17, 1]$. Therefore, we can outline the following cases:

- If $\delta \in [0, 9/17)$, firms play *à la* Cournot-Nash forever. Hence, at $t = 0$, their technological arrangements are driven by the following chain of inequalities:

  $$\pi_H^{CN}(IV) > \pi^{CN}(RJV) > \pi_L^{CN}(IV) \quad (8.49)$$

  which suffices to establish that firms will activate independent ventures due to the profit incentive characterising firm H. Note that:

  $$2\pi^{CN}(RJV) > \pi_H^{CN}(IV) + \pi_L^{CN}(IV) \quad (8.50)$$

  so that neither the redistribution of profits from firm L to firm H after activating the joint venture would solve the problem.

- If $\delta \in [9/17, 1]$, Lemma 3 holds. Firms collude in output levels forever; given this, it is clear that $RJV \succ IV$ for both of them, since $\pi^{cart}(RJV) = 2\pi^{cart}(IV)$.

8.6.2 Venture commitments in the Bertrand setting

Here, the following taxonomy applies:

- If $\delta \in [0, 1/2)$, firms play *à la* Bertrand-Nash forever. The organizational arrangement of the R&D investment depends upon whether firms may credibly commit to redistribute profits or not. To see this, observe that

$$\pi_H^{BN}(IV) > \pi^{BN}(RJV) = \frac{\pi^M}{4} > \pi_L^{BN}(IV). \tag{8.51}$$

However:

$$2\pi^{BN}(RJV) = \frac{\pi^M}{2} > \pi_H^{BN}(IV) + \pi_L^{BN}(IV) \tag{8.52}$$

and a (rather considerable) profit redistribution from firm L to firm H, thereby reducing the share of profits of the former following the RJV would allow both firms to activate a joint venture. However, this possibility, as is well known, is less than credible, and therefore is not taken seriously, in general.

- If $\delta \in [1/2, 1]$, Lemma 3 holds. Firms collude in prices forever; then, clearly, $RJV \succ IV$ for both of them, since $\pi^{cart}(RJV) = 2\pi^{cart}(IV)$.

The foregoing discussion highlights that, in both settings, either firms *cooperate* in R&D *and collude* in the relevant market variable, or they *activate independent ventures and compete* on the market place. Contrary to some of the existing literature dealing with the same issue in other models (Lambertini, Poddar and Sasaki, 1998; 2002; 2003), here there are no nuances, as synthesised in the following:[10]

Proposition 8.1 *For all $\delta \in [0, \delta^I)$, $I = B, C$, firms choose independent ventures and compete in the market supergame. For all $\delta \in [\delta^I, 1]$, firms undertake a joint venture and collude in the market supergame.*

This result confirms the conclusions drawn by Martin (1995) using a Cournot model with homogeneous goods, where firms evaluate the choice between IV and RJV for a process innovation. The present analysis appears to imply that the risk of undesirable connections between technological cooperation and price or quantity collusion is also present when firms are searching for a product innovation.

Moreover, it is worth stressing that firms could switch from Cournot to Bertrand behaviour for all $\delta \in [1/2, 9/17)$, wherein collusion can be sustained

[10] This statement of course disregards the possibility that firms redistribute profits in the case of a joint venture followed by non-cooperative price setting behaviour.

in prices but not in quantities, while ensuring the same discounted profit flows. Moreover, whenever firms are unable to collude along the frontier of monopoly profits, they could well aim at any partial collusion compatible with their intertemporal preferences. This perspective is somewhat intriguing but also less tractable.

Finally, it can also be pointed out that the type of price/quantity behaviour (collusive or non-cooperative), jointly with the venture decision taken by firms, ultimately affects the equilibrium quality levels. A straightforward comparison allows us to draw the following conclusive result:

Proposition 8.2 *Undertaking a joint venture in the R&D phase, followed by collusive behaviour in the market supergame allows firms to substantially increase the equilibrium quality level. This holds irrespective of whether firms are Bertrand or Cournot players.*

This result has an intuitive explanation, as both collusion (in prices or in quantities) and an *RJV* have a positive effect on firms' profits and consequently on their ability to invest more (jointly) in order to enhance the quality of the good supplied to consumers.

8.7 Concluding remarks

We have reassessed an issue previously raised by Martin (1995), under a new perspective, where firms' initial R&D efforts are aimed at product rather than process innovation. We have analysed the relationship between the organizational design of R&D for product innovation and the stability of implicit collusion either in quantities or in prices, keeping unaltered w.r.t. Martin's paper the rules governing the market supergame, i.e., using the Perfect Folk Theorem.

The main conclusion emerging from this setting is that an *RJV* drastically contributes to stabilise implicit collusion, independently of the market variable being set by firms. Putting it differently, the foregoing analysis suggests that, whenever firms consider it profitable to undertake a joint venture for product innovation, the latter consisting in quality improvement, then it should be taken as a strong hint that these firms may subsequently collude in the market phase. Therefore, public policies towards R&D behaviour should be carefully designed so as not to become inconsistent with the pro-competitive attitude characterising the current legislation on marketing practices. There remains, of course, to be investigated the possibility of partial collusion. This is left for future research.

References

1] Abreu, D. (1986), "Extremal Equilibria of Oligopolistic Supergames", *Journal of Economic Theory*, **39**, 191-225.
2] Abreu, D. (1988), "On the Theory of Infinitely Repeated Games with Discounting", *Econometrica*, **56**, 383-96.
3] Abreu, D.J., D. Pearce and E. Stacchetti (1986), "Optimal Cartel Equilibria with Imperfect Monitoring", *Journal of Economic Theory*, **39**, 251-69.
4] Albæk, S. and L. Lambertini (1998), "Collusion in Differentiated Duopolies Revisited", *Economics Letters*, **59**, 305-8.
5] Aoki, R. and T. Prusa (1997), "Sequential versus Simultaneous Choice with Endogenous Quality", *International Journal of Industrial Organization*, **15**, 103-21.
6] Beath, J., Y. Katsoulacos and D. Ulph (1987), "Sequential Product Innovation and Industry Evolution", *Economic Journal*, **97**, 32-43.
7] Cabral, L.M.B. (1996), "R&D Alliances as Non-Cooperative Supergames", CEPR Discussion Paper no. 1439.
8] Cabral, L.M.B. (2000), "R&D Cooperation and Product Market Competition", *International Journal of Industrial Organization*, **18**, 1033-47.
9] Chang M. (1991), "The Effects of Product Differentiation on Collusive Pricing", *International Journal of Industrial Organization*, **9**, 453-69.
10] Chang, M.H. (1992), "Intertemporal Product Choice and Its Effects on Collusive Firm Behavior", *International Economic Review*, **33**, 773-93.
11] Choi, C.J. and H.S. Shin (1992), "A Comment on a Model of Vertical Product Differentiation", *Journal of Industrial Economics*, **40**, 229-31.
12] Deneckere, R. (1983), "Duopoly Supergames with Product Differentiation", *Economics Letters*, **11**, 37-42.
13] Dutta, P.K., S. Lach and A. Rustichini (1995), "Better Late than Early: Vertical Differentiation in the Adoption of a New Technology", *Journal of Economics and Management Strategy*, **4**, 563-89.
14] EC Commission (1990), *Competition Law in the European Communities*, Volume I, *Rules Applicable to Undertakings*, Brussels-Luxembourg, EC Commission.
15] Friedman, J.W. (1971), "A Noncooperative Equilibrium for Supergames", *Review of Economic Studies*, **38**, 1-12.
16] Friedman, J.W. and J.-F. Thisse (1993), "Partial Collusion Fosters Minimum Product Differentiation", *RAND Journal of Economics*, **24**, 631-45.
17] Fudenberg, D. and E. Maskin (1986), "The Folk Theorem in Repeated Games with Discounting or with Incomplete Information", *Econometrica*, **54**, 533-54.

18] Gabszewicz, J.J. and J.-F. Thisse (1979), "Price Competition, Quality and Income Disparities", *Journal of Economic Theory*, **20**, 340-59.
19] Gabszewicz, J.J. and J.-F. Thisse (1980), "Entry (and Exit) in a Differentiated Industry", *Journal of Economic Theory*, **22**, 327-38.
20] Goto, A. and R. Wakasugi (1988), "Technology Policy", in Komiya, R., M. Okuno and K. Suzumura (eds.), *Industrial Policy of Japan*, New York, Academic Press.
21] Häckner, J. (1994), "Collusive Pricing in Markets for Vertically Differentiated Products", *International Journal of Industrial Organization*, **12**, 155-77.
22] Häckner, J. (1995), "Endogenous Product Design in an Infinitely Repeated Game", *International Journal of Industrial Organization*, **13**, 277-99.
23] Häckner, J. (1996), "Optimal Symmetric Punishments in a Bertrand Differentiated Products Duopoly", *International Journal of Industrial Organization*, **14**, 611-30.
24] Lambertini, L. (1996), "Cartel Stability and the Curvature of Market Demand", *Bulletin of Economic Research*, **48**, 329-34.
25] Lambertini, L. (1997), "Prisoners' Dilemma in Duopoly (Super)Games", *Journal of Economic Theory*, **77**, 181-91.
26] Lambertini, L. (1999), "Endogenous Timing and the Choice of Quality in a Vertically Differentiated Duopoly", *Research in Economics* (*Ricerche Economiche*), **53**, 101-09.
27] Lambertini, L. (2000), "Technology and Cartel Stability under Vertical Differentiation", *German Economic Review*, **1**, 421-42.
28] Lambertini, L. and D. Sasaki (1999), "Optimal Punishments in Linear Duopoly Supergames with Product Differentiation", *Journal of Economics*, **69**, 173-88.
29] Lambertini, L. and D. Sasaki (2002), "Non-Negative Quantity Constraints and the Duration of Punishment", *Japanese Economic Review*, **53**, 77-93.
30] Lambertini, L., S. Poddar and D. Sasaki (1998), "Standardization and the Stability of Collusion", *Economics Letters*, **58**, 303-10.
31] Lambertini, L., S. Poddar and D. Sasaki (2002), "Research Joint Ventures, Product Differentiation, and Price Collusion", *International Journal of Industrial Organization*, **20**, 829-54.
32] Lambertini, L., S. Poddar and D. Sasaki (2003), "RJVs in Product Innovation and Cartel Stability", *Review of Economic Design*, **7**, 465-77.
33] Lehmann-Grube, U. (1997), "Strategic Choice of Quality when Quality is Costly: The Persistence of the High-Quality Advantage", *RAND Journal of Economics*, **28**, 372-84.
34] Majerus, D.W. (1988), "Price vs Quantity Competition in Oligopoly Supergames", *Economics Letters*, **27**, 293-7.

35] Martin, S. (1995), "R&D Joint Ventures and Tacit Product Market Collusion", *European Journal of Political Economy*, **11**, 733-41.
36] Motta, M. (1992), "Cooperative R&D and Vertical Product Differentiation", *International Journal of Industrial Organization*, **10**, 643-61.
37] Motta, M. (1993), "Endogenous Quality Choice: Price vs Quantity Competition", *Journal of Industrial Economics*, **41**, 113-32.
38] Rosenkranz, S. (1995), "Innovation and Cooperation under Vertical Product Differentiation", *International Journal of Industrial Organization*, **13**, 1-22.
39] Rosenkranz, S. (1997), "Quality Improvements and the Incentive to Leapfrog", *International Journal of Industrial Organization*, **15**, 243-61.
40] Ross, T.W. (1992), "Cartel Stability and Product Differentiation", *International Journal of Industrial Organization*, **10**, 1-13.
41] Rothschild, R. (1992), "On the Sustainability of Collusion in Differentiated Duopolies", *Economics Letters*, **40**, 33-7.
42] Shaked, A. and J. Sutton (1982), "Relaxing Price Competition through Product Differentiation", *Review of Economic Studies*, **49**, 3-13.
43] Shaked, A. and J. Sutton (1983), "Natural Oligopolies", *Econometrica*, **51**, 1469-83.
44] Singh, N. and X. Vives (1984), "Price and Quantity Competition in a Differentiated Duopoly", *RAND Journal of Economics*, **15**, 546-54.
45] van Dijk, T. (1996), "Patent Height and Competition in Product Improvements", *Journal of Industrial Economics*, **44**, 151-67.

Chapter 9

Product innovation with sequential entry

Giorgia Bertuzzi and Luca Lambertini

9.1 Introduction

The existing literature on vertically differentiated markets maintains that superior qualities command higher prices, market shares, and profits than inferior qualities. Accordingly, it is widely accepted that earlier entrants will strive to fill the most profitable market niches by supplying high-quality goods, leaving the task of supplying consumers located in the lower part of the income distribution to later entrants (Gabszewicz and Thisse, 1979, 1980; Shaked and Sutton, 1982, 1983). In this literature, however, Nash (i.e., simultaneous) behaviour is generally assumed on the part of firms. That is, the realistic feature of sequential innovation and entry is not explicitly treated.[1]

Notwithstanding the relevance of the topic, the issue of product innovation has been relatively little treated in relation to vertical differentiation. The few existing contributions (Beath, Katsoulacos and Ulph, 1987; Dutta, Lach and Rustichini, 1995; Rosenkranz, 1995, 1997; van Dijk, 1996) stress once again the importance of being the high-quality firm. This, to a large extent, is in contrast with the empirical evidence, or just casual observation. E.g., it is surely more profitable to sell Honda Hornets rather than Ducati

[1]Sequential entry in Stackelberg games has been investigated by Aoki and Prusa (1997); Lehmann-Grube (1997) and Lambertini (1999). Also these contributions, however, find that 'earlier entrants' should supply high-quality goods, due to the static approach adopted. See also Donnenfeld and Weber (1992, 1995).

749s, although everybody may agree that the former is characterised by a lower quality than the latter. The reasons are that (i) the relatively lower quality may take less to develop and introduce into the market, and therefore it may enjoy monopoly profits for some time; (ii) a lower quality may attract a larger demand, although it commands a lower price, than the higher one. If the quantity effect dominates the price effect, then a low-quality good turns out to be more profitable than a superior alternative.

Our purpose is to investigate the issue of product innovation and sequential quality choice in a vertically differentiated market under the assumption of full market coverage (holding irrespective of whether the market is a monopoly or a duopoly). The game unravels over an infinite horizon, and two firms enter the market at different points in time, with goods characterised by different quality levels. In particular, the first entrant's quality is lower than the second entrant's. This feature reflects the idea that quality improvements take time, and a firm has to trade off an earlier entry against an additional quality increase. We shall take into consideration two approaches. In the first, product innovation is deterministic, with the innovation taking place at an exogenously given date. In the second, innovation is stochastic and occurs with a fixed probability. We show that, in both cases, the first entrant strictly prefers to supply the inferior quality since this ensures higher profits as compared to the situation where the same firm offers a high-quality good.

The remainder of the chapter is organised as follows. The model is laid out in section 9.2. Sections 9.3 and 9.4 deal with deterministic and stochastic innovation, respectively. The common features of the two approaches are briefly discussed in section 9.5. Concluding comments are in section 9.6.

9.2 Setup and price behaviour

The basic model shares many features with Cremer and Thisse (1991, 1994), Lambertini (1996) and several others concerning the demand and cost structure, and Lambertini (2002) concerning the structure of the innovation race. Consider a market for vertically differentiated products where consumers are indexed by their respective marginal willingness for quality, measured by parameter $\theta \in \left[\underline{\theta}, \overline{\theta}\right]$, with $\underline{\theta} = \overline{\theta} - 1$. The distribution of θ over the support $\left[\underline{\theta}, \overline{\theta}\right]$ is uniform, with unit density, so that the total mass of consumer is normalised to one.

The market exists over $t \in [0, \infty)$. Time is treated as a continuous variable. Two profit-maximising single-product firms, labelled as H and L, sequentially choose qualities $q_H \geq q_L \geq 0$, and simultaneously compete in prices as soon as both are in the market (if ever). Throughout the game, both firms exhibit the same constant discount rate $\rho > 0$. Product quality must be chosen once and for all at the time of entry so as to maximise the discounted flow of (expected) profits from that instant to doomsday. Unit production cost is assumed to be constant for each variety, and equal to cq_i^2, $i = H, L$, where c is a positive parameter. Hence, the instantaneous total costs borne by firm i are:

$$C_i = cq_i^2 x_i \tag{9.1}$$

where x_i is the instantaneous output level produced by firm i at any time t during the game.

We assume that the market is fully covered irrespective of whether it is supplied by one or two firms. Each consumer draws the net utility:

$$U = \theta q_i - p_i \geq 0 \tag{9.2}$$

from the consumption of one unit of good i in each period, paying the price p_i. If both firms operate in the market, their respective demands are:

$$x_H = \overline{\theta} - \frac{p_H - p_L}{q_H - q_L}\,;\; x_L = 1 - x_H\,. \tag{9.3}$$

The first product is introduced at time $t = 0$, with certainty. The second variety enters at the date $\tau \in [0, \infty)$. We shall alternatively assume either:

Assumption 9.1 *τ is certain and* a priori *known to both firms.*

or:

Assumption 9.2 *τ is uncertain and* a priori *unknown to both firms. Entry may occur at any $\tau \in [0, \infty)$ with a fixed instantaneous probability $\mathfrak{p} \in [0, 1]$.*

As we are going to verify in the remainder of the analysis, the model resulting from Assumption 1 is the *certainty-equivalent* version of the model resulting from Assumption 2.

In either case, the instantaneous profits are easily obtained, under both monopoly and duopoly. Over $t \in [0, \tau)$, the first entrant is a monopolist, and, given the full market coverage assumption, sets price so as to drive to zero the net utility of the marginal consumer indexed by $\underline{\theta} = \overline{\theta} - 1$:

$$p^M = \left(\overline{\theta} - 1\right) q. \tag{9.4}$$

Therefore, instantaneous monopoly profits write:

$$\pi^M = \left(p^M - cq^2\right) x^M = \left[\left(\overline{\theta} - 1\right) - cq\right] q \tag{9.5}$$

since $x^M = 1$. Profits π^M are concave and single-peaked in q, with the maximum at $q^M = \left(\overline{\theta} - 1\right)/(2c)$, which is the unconstrained optimal monopoly quality: the monopolist serving all consumers supplies the quality preferred by the poorest customer in the market.[2] This, in general, is obviously not optimal in the present model, unless the first entrant remains a monopolist forever (i.e., $h = 0$). Notice, however, that this suggests that the first entrant will locate in the lower part of the quality spectrum, with the second innovator filling eventually the resulting gap in the upper part of the product range. Accordingly, we label the first entrant as firm L and the second entrant as firm H.

The instantaneous profits of the two firms over $t \in [\tau, \infty)$ are:

$$\begin{aligned} \pi_H &= \left(p_H^D - cq_H^2\right) x_H^D = \left(p_H^D - cq_H^2\right)\left(\overline{\theta} - \frac{p_H - p_L}{q_H - q_L}\right) \\ \pi_L &= \left(p_L^D - cq_L^2\right) x_L^D = \left(p_L^D - cq_L^2\right)\left(1 - \overline{\theta} + \frac{p_H - p_L}{q_H - q_L}\right) \end{aligned} \tag{9.6}$$

where superscript D stands for *duopoly*. The instantaneous profits accruing to the first entrant in the *ad interim* monopoly phase can be rewritten as $\pi_L^M = \left[\left(\overline{\theta} - 1\right) - cq_L\right] q_L$.

In solving the duopoly game for $t \in [\tau, \infty)$, we assume that firm L acts as a Stackelberg leader in the quality stage, while pricing behaviour is taken to be simultaneous. As a first step, we can quickly solve the price stage by checking that

$$\begin{aligned} p_H &= \frac{(q_H - q_L)(\overline{\theta} + 1) + 2cq_H^2 + cq_L^2}{3} \\ p_L &= \frac{(q_H - q_L)(2 - \overline{\theta}) + 2cq_L^2 + cq_H^2}{3} \end{aligned} \tag{9.7}$$

are the Bertrand-Nash equilibrium prices for any quality pair (q_H, q_L) (see Cremer and Thisse, 1994; and Lambertini, 1996). Plugging (9.7) into (9.6) and rearranging, we get the relevant per-period profits defined in terms of quality levels only:

$$\begin{aligned} \pi_H^D &= \frac{(q_H - q_L)\left[\overline{\theta} + 1 - c(q_H + q_L)\right]^2}{9} \\ \pi_L^D &= \frac{(q_H - q_L)\left[2 - \overline{\theta} + c(q_H + q_L)\right]^2}{9} \end{aligned} \tag{9.8}$$

[2] For further details, see Lambertini (1997).

Now we can solve the two alternative entry (or innovation) games, starting with the analysis of the case where the second innovation is introduced with certainty at time τ.

9.3 Game I: Certain innovation

In this setting, firm L is a monopolist over $t \in [0,\tau)$, while Stackelberg competition takes place in qualities over $t \in [\tau,\infty)$, with firm L leading. The objective of firm H (the follower) is:

$$\max_{q_H} \Pi_H = \int_\tau^\infty e^{-\rho t}\pi_H^D dt = \frac{(q_H - q_L)\left[\overline{\theta} + 1 - c(q_H + q_L)\right]^2}{9\rho e^{\rho\tau}} \tag{9.9}$$

while the objective of firm L (the leader) is:

$$\max_{q_L} \Pi_L = \int_0^\tau e^{-\rho t}\pi_L^M dt + \int_\tau^\infty e^{-\rho t}\pi_L^D dt \tag{9.10}$$

$$= \frac{9(1-e^{\rho\tau})\left[(\overline{\theta}-1) - cq_L\right]q_L + (q_H - q_L)\left[2 - \overline{\theta} + c(q_H + q_L)\right]^2}{9\rho e^{\rho\tau}}$$

under the constraint given by the follower's reaction function:[3]

$$\frac{\partial \Pi_H}{\partial q_H} = 0 \Rightarrow q_H^* = \frac{\overline{\theta} + 1 + cq_L}{3c}. \tag{9.11}$$

The first order condition of the leader is:

$$\frac{\partial \Pi_L}{\partial q_L} = \left[67 - 81e^{\rho\tau} + \overline{\theta}\left(81e^{\rho\tau} + 8\overline{\theta} - 49 + 32cq_L\right) + \right. \tag{9.12}$$

$$\left. -2cq_L\left(81e^{\rho\tau} - 49 + 16cq_L\right)\right] / \left(81\rho e^{\rho\tau}\right) = 0$$

yielding two solutions:

$$q_{L1} = \frac{16\overline{\theta} + 49 - 81e^{\rho\tau} - 3\sqrt{505 + 9e^{\rho\tau}\left(81e^{\rho\tau} - 130\right)}}{32c};$$
$$q_{L2} = \frac{16\overline{\theta} + 49 - 81e^{\rho\tau} + 3\sqrt{505 + 9e^{\rho\tau}\left(81e^{\rho\tau} - 130\right)}}{32c}. \tag{9.13}$$

[3] There exists another solution to the follower's first order condition for profit maximisation. However, it can be excluded since it does not meet the second order conditions.

In order to determine which one of the two critical points identified the leader's optimum, one can resort to the second order conditions. Alternatively, by taking the limit of both roots in (9.13) for $\tau \to 0$, one finds:

$$\lim_{\tau \to 0} q_{L1} = \frac{2\overline{\theta} - 7}{4c} \, ; \, \lim_{\tau \to 0} q_{L2} = \frac{2\overline{\theta} - 1}{4c} \, . \tag{9.14}$$

The latter coincides with the leader's optimal quality in the static game (see Lambertini, 1996). The same conclusion would obviously hold by virtue of the concavity conditions. Therefore, q_{L2} qualifies as the optimal quality level for the first entrant.

Now it is worth noting that the statically optimal quality of the leader coincides with the quality preferred by the average (and median) consumer. On the other hand, it can also be verified that, should the first entrant stand alone in the market forever, i.e., $\tau \to \infty$, it would obviously produce $q_L = q^M$ forever. It is trivial to check that $\partial q_{L2}/\partial \tau < 0$ everywhere. Hence, without further proof, we can state the following:

Proposition 9.1 *In the deterministic innovation game, the leader's optimal quality is*

$$q_L^* = \frac{16\overline{\theta} + 49 - 81e^{\rho\tau} + 3\sqrt{505 + 9e^{\rho\tau}\left(81e^{\rho\tau} - 130\right)}}{32c}$$

with

$$q_L^* \in \left(\frac{\overline{\theta} - 1}{2c} \, ; \, \frac{2\overline{\theta} - 1}{4c} \right] \text{ for all } \tau \in [0, \infty) \, .$$

The foregoing analysis also permits us to draw an interesting conclusion about the profit ranking. To avoid boring calculations, just observe that, if $\tau \to 0$, the game replicates a static Stackelberg game at any t. In such a case, instantaneous profits are (see Lambertini, 1996):

$$\pi_H^D = \frac{1}{18c} < \pi_L^D = \frac{2}{9c} \tag{9.15}$$

while if $\tau \to \infty$, the first entrant (firm L) remains a monopolist forever. In these two extreme situations, the low-quality firm obtains higher profits than the high-quality firm. The same holds for any $\tau \in (0, \infty)$, in which case the low-quality firm enjoys some *ad interim* monopoly power, followed by the quality leadership in the duopoly phase. Therefore, we can formulate:

Proposition 9.2 *In the deterministic game, the first entrant supplies the low-quality product, but gains higher profits than the second innovator.*

9.4 Game II: Innovation under uncertainty

To model the case of stochastic innovation, we consider that the second entrant may introduce the new product q_H with probability $\mathfrak{p} \in [0,1]$. Then, the expected flow of discounted profits for firm L writes as follows:

$$E(\Pi_L) = \int_0^{\infty} \left[(1-\mathfrak{p})\,\pi_L^M + \mathfrak{p}\pi_L^D\right] e^{-(\rho+\mathfrak{h})t} dt. \tag{9.16}$$

If $\mathfrak{p} = 0$, firm H never enters the market, and firm L remains a monopolist forever, while if $\mathfrak{p} = 1$ firm H enters immediately at $t = 0$ and the market is a duopoly from the outset. The leader's problem consists in:

$$\begin{array}{c} \max_{q_L} E(\Pi_L) = \int_0^{\infty} \left[(1-\mathfrak{p})\,\pi_L^M + \mathfrak{p}\pi_L^D\right] e^{-(\rho+\mathfrak{h})t} dt \\ s.t.: \; q_H^* = \dfrac{\overline{\theta}+1+cq_L}{3c}. \end{array} \tag{9.17}$$

Observe that, obviously, the constraint given by the follower's best reply function remains the same as in the deterministic case. Simplifying $E(\Pi_L)$, we obtain the following expression:

$$E(\Pi_L) = \frac{\mathfrak{p}\left(1+\overline{\theta}-2cq_L\right)\left(7-2\overline{\theta}+4cq_L\right)^2 - q_L\left(1-\overline{\theta}+cq_L\right)(1-\mathfrak{p})}{243c\,(\rho+\mathfrak{p})}. \tag{9.18}$$

From the first order condition w.r.t. q_L we find two candidate solutions to the constrained optimum problem:

$$\begin{aligned} q_{L1} &= \frac{\left(16\overline{\theta}+49\right)\mathfrak{p} - 81 - 3\sqrt{505\mathfrak{p}^2 + 9\,(81-130\mathfrak{p})}}{32c}; \\ q_{L2} &= \frac{\left(16\overline{\theta}+49\right)\mathfrak{p} - 81 + 3\sqrt{505\mathfrak{p}^2 + 9\,(81-130\mathfrak{p})}}{32c}. \end{aligned} \tag{9.19}$$

Then, by checking either the second order condition, or the limits:

$$\lim_{\mathfrak{p}\to 1} q_{L1} = \frac{2\overline{\theta}-7}{4c}; \; \lim_{\mathfrak{p}\to 1} q_{L2} = \frac{2\overline{\theta}-1}{4c}, \tag{9.20}$$

we can conclude that the maximum expected profits are attained by the leader at q_{L2}. As in the deterministic case, the statically optimal quality of the leader (when $\mathfrak{p} = 1$) coincides with the quality preferred by the average (and median) consumer. Moreover, if the first entrant were to remain alone in the market forever (i.e., $\mathfrak{p} = 0$), it would supply $q_L = q^M$ at any t. It is also simple to check that $\partial q_{L2}/\partial\mathfrak{p} > 0$ everywhere. Therefore, the following holds:

Proposition 9.3 *In the stochastic innovation game, the leader's optimal quality is*

$$q_L^* = \frac{(16\overline{\theta} + 49)\,\mathfrak{p} - 81 + 3\sqrt{505\mathfrak{p}^2 + 9(81 - 130\mathfrak{p})}}{32c}$$

with

$$q_L^* \in \left(\frac{\overline{\theta} - 1}{2c}, \frac{2\overline{\theta} - 1}{4c}\right] \text{ for all } \mathfrak{p} \in [0, 1].$$

By plugging q_L^* into the follower's best reply function and rearranging, we find firm H's optimal quality:

$$q_H^* = \frac{16\overline{\theta}\mathfrak{p} - 27(\mathfrak{p} - 1) + \sqrt{505\mathfrak{p}^2 + 9(81 - 130\mathfrak{p})}}{32c\mathfrak{p}}. \qquad (9.21)$$

Concerning expected equilibrium profits, we can compare $E(\Pi_L)$ and $E(\Pi_H)$ at $\mathfrak{p} = 0$ and $\mathfrak{p} = 1$, alternatively, to find that in both cases the leader's profits are larger than the follower's. The same obviously holds for any intermediate length of the *ad interim* monopoly phase. Accordingly, we can formulate:

Proposition 9.4 *In the stochastic game, the first entrant supplies the low-quality product, but gains higher expected profits than the second innovator.*

9.5 Comments and extensions

On the basis of (9.13) and (9.19), one can easily verify that the optimal quality levels of the leader in the two settings are the same, up to the following transformation:

$$\tau = \frac{\log(1/\mathfrak{p})}{\rho}. \qquad (9.22)$$

The same consideration obviously applies to the follower's quality choice as well as equilibrium profits in the two alternative approaches. An alternative formulation of the stochastic innovation model, preserving this property, is that where $\mathfrak{h}$ is the *hazard rate* (i.e., the constant probability that, at any t, the follower may enter the market),[4] with $\mathfrak{h} \in [0, \infty)$. Here, the objective function of the leader is:

$$E(\Pi_L) = \int_0^{\infty} \left[\pi_L^M + \mathfrak{h}\pi_L^D\right] e^{-(\rho+\mathfrak{h})t} dt = \frac{\pi_L^M + \mathfrak{h}\pi_L^D}{\rho + \mathfrak{h}}. \qquad (9.23)$$

[4]This approach, which assumes that the innovation follows a Poisson process, is widely adopted in the existing literature on stochastic R&D. For an exhaustive survey, see Reinganum (1989).

If $\mathfrak{h} = 0$, firm L remains a monopolist forever, while if $\mathfrak{h} \to \infty$ firm L is a Stackelberg leader in the choice of quality with the follower entering at the outset. In this case, moving from the stochastic approach to the deterministic one requires the following transformation:

$$\mathfrak{h} = \frac{1}{e^{\rho\tau} - 1}. \tag{9.24}$$

The foregoing analysis could be extended to allow for partial market coverage during the monopoly phase, and possibly also the duopoly phase. Moreover, other cost structures could be considered. For instance, the cost function might be additively separable in output and quality levels. At this stage, the consequences of modifying these two assumptions are largely unpredictable.

9.6 Concluding remarks

We have examined the product innovation process taking place in a vertically differentiated industry under the assumption of full market coverage. Alternatively, we have considered the cases of deterministic and stochastic innovation, showing that, in both cases, the first entrant strictly prefers to supply the inferior quality since this ensures higher profits as compared to the situation where the same firm offers a high-quality good. These findings are in sharp contrast with those emerging from the previous literature on the same topic, where typically the time dimension is not explicitly accounted for.

References

1] Aoki, R. and T. Prusa (1997), "Sequential versus Simultaneous Choice with Endogenous Quality", *International Journal of Industrial Organization*, **15**, 103-21.
2] Beath, J., Y. Katsoulacos and D. Ulph (1987), "Sequential Product Innovation and Industry Evolution", *Economic Journal*, **97**, 32-43.
3] Cremer, H. and J.-F. Thisse (1991), "Location Models of Horizontal Differentiation: A Special Case of Vertical Differentiation Models", *Journal of Industrial Economics*, **39**, 383-90.
4] Cremer, H. and J.-F. Thisse (1994), "Commodity Taxation in a Differentiated Oligopoly", *International Economic Review*, **35**, 613-33.

5] Donnenfeld, S. and S. Weber (1992), "Vertical Product Differentiation with Entry", *International Journal of Industrial Organization*, **10**, 449-72.
6] Donnenfeld, S. and S. Weber (1995), "Limit Qualities and Entry Deterrence", *RAND Journal of Economics*, **26**, 113-30.
7] Dutta, P.K., S. Lach and A. Rustichini (1995), "Better Late than Early: Vertical Differentiation in the Adoption of a New Technology", *Journal of Economics and Management Strategy*, **4**, 563-89.
8] Gabszewicz, J.J. and J.-F. Thisse (1979), "Price Competition, Quality and Income Disparities", *Journal of Economic Theory*, **20**, 340-59.
9] Gabszewicz, J.J. and J.-F. Thisse (1980), "Entry (and Exit) in a Differentiated Industry", *Journal of Economic Theory*, **22**, 327-38.
10] Lambertini, L. (1996), "Choosing Roles in a Duopoly for Endogenously Differentiated Products", *Australian Economic Papers*, **35**, 205-24.
11] Lambertini, L. (1997), "The Multiproduct Monopolist under Vertical Differentiation: An Inductive Approach", *Recherches Economiques de Louvain*, **63**, 109-22.
12] Lambertini, L. (1999), "Endogenous Timing and the Choice of Quality in a Vertically Differentiated Duopoly", *Research in Economics* (*Ricerche Economiche*), **53**, 101-09.
13] Lambertini, L. (2002), "Equilibrium Locations in a Spatial Model with Sequential Entry in Real Time", *Regional Science and Urban Economics*, **32**, 47-58.
14] Lehmann-Grube, U. (1997), "Strategic Choice of Quality when Quality is Costly: The Persistence of the High-Quality Advantage", *RAND Journal of Economics*, **28**, 372-84.
15] Reinganum, J.F. (1989), "The Timing of Innovation: Research, Development, and Diffusion", in Schmalensee, R. and R. Willig (eds), *Handbook of Industrial Organization*, Vol. 1., Amsterdam, North-Holland, 849-908.
16] Rosenkranz, S. (1995), "Innovation and Cooperation under Vertical Product Differentiation", *International Journal of Industrial Organization*, **13**, 1-22.
17] Rosenkranz, S. (1997), "Quality Improvements and the Incentive to Leapfrog", *International Journal of Industrial Organization*, **15**, 243-61.
18] Shaked, A. and J. Sutton (1982), "Relaxing Price Competition through Product Differentiation", *Review of Economic Studies*, **49**, 3-13.
19] Shaked, A. and J. Sutton (1983), "Natural Oligopolies", *Econometrica*, **51**, 1469-83.
20] van Dijk, T. (1996), "Patent Height and Competition in Product Improvements", *Journal of Industrial Economics*, **44**, 151-67.

Chapter 10

A differential game with quality improvement

Luca Lambertini

10.1 Introduction

Here I propose a dynamic approach to the strategic use of non-price tools in a differential game model of vertical differentiation. The applications of differential games to economic analysis have remarkably increased over the last two decades, to deal with several different topics, including oligopoly theory.[1]

Ever since the pioneering work of Spence (1975) and Mussa and Rosen (1978) on the provision of product quality by a monopolist, vertical differentiation has received wide attention within the theory of industrial organization. Several issues have been investigated in oligopoly models where firms supply goods of different quality. In Gabszewicz and Thisse (1979, 1980) and Shaked and Sutton (1982, 1983), the so-called *finiteness property* is established, according to which the number of firms that can survive in a vertically differentiated market is finite. This result holds if unit costs of quality are flat enough, and the overall cost associated with the improvement of quality is an R&D cost unrelated with the scale of production. In their approach, the only cost explicitly modelled is a fixed cost which is assumed to be exogenous and arbitrarily small. Therefore, the finiteness property essentially depends on demand rather than technological conditions. The influence of the shape of the cost function on prices, market shares and profits is the topic of several

[1] For an overview, see Clemhout and Wan (1994), Dockner *et al.* (2000) and Cellini and Lambertini (2003a).

contributions, where the cost of quality is alternatively related or unrelated with the output scale.[2]

More recent contributions deal with several aspects of the technology associated with product innovation in vertically differentiated markets, through either independent ventures (Beath *et al.*, 1987; Dutta *et al.*, 1995; Rosenkranz, 1997) or joint ventures (Motta, 1992; Rosenkranz, 1995; Lambertini, 2000; Lambertini *et al.*, 2002). A result common to all these contributions is that the highest quality good is more profitable than all inferior varieties, irrespective of the specification of the cost function and, in particular, notwithstanding the assumption, common to all this literature, that the higher the quality of a good, the higher its cost.

With the exception of Beath *et al.* (1987) and Dutta *et al.* (1995), where quality improvement is modelled as the outcome of an uncertain innovation race, the above literature adopts a static approach where firms set qualities and prices (or outputs) in two stages. To the best of my knowledge, the problem of quality supply has been investigated in optimal control and differential game models only in relation with advertising strategies designed to increase goodwill and/or to compete for market shares (Kotowitz and Mathewson, 1979; Conrad, 1985; and Ringbeck, 1985).[3]

I investigate a differential duopoly game where firms supply goods of different quality, which is the result of capital accumulation over time. The degree of vertical (quality) differentiation interacts with prices in determining market shares at every instant. The setup of market demand is borrowed from well known static models, while the evolution of qualities over time is modelled according to two different dynamic equations describing two alternative approaches to R&D activity. In the first, instantaneous costs are linear in the quality level, while the dynamics of quality are characterised by decreasing returns to R&D efforts. In the second, the opposite holds. The introduction of dynamic capital accumulation allows me to show that, on the one hand, several results of the static approach are not robust. First, the sustainability of the duopoly regime depends upon the non-negativity of firms' profits, which in turn depends on the size of their respective investment in R&D to improve quality. Second, the dynamic model produces situations

[2] For models where the development of quality bears upon variable costs, see Moorthy (1988); Champsaur and Rochet (1989); Cremer and Thisse (1994); Lambertini (1996). For those where quality represents a fixed cost, see Aoki and Prusa (1997); Lehmann-Grube (1997) and Lambertini (1999). A comparative evaluation is in Motta (1993).

[3] For exhaustive surveys on dynamic advertising, see Sethi (1977); Jørgensen (1982); Feichtinger and Jørgensen (1983); Erickson (1991); Feichtinger *et al.* (1994). For oligopoly models with dynamic advertising, see Leitmann and Schmitendorf (1978), Feichtinger (1983) and Cellini and Lambertini (2003b).

where the low quality firm earns higher profits than the high quality firm.[4] On the other hand, the dynamic results reinforce some of the wisdom we are accustomed to from the static literature. One such result is that the high quality firm always invests more than the low-quality firm. The second is that, in both models investigated here, the optimal quality ratio is 4/7, i.e., the same as in the static model with zero production costs, as in Choi and Shin (1992). When dealing with the static literature on product differentiation, one may consider the case of costless quality improvements as a theoretical *curiosum* or a simplifying assumption towards the analysis of equilibrium market structure as described by the finiteness property. The dynamic analysis presented here seems to point out that the so-called *4/7 rule* is more than that, since the static two-stage model of Choi and Shin summarises the essential steady-state properties of a dynamic approach to the issue of vertical differentiation. When comparing static and differential games, this appears to be the exception rather than the rule.[5]

The remainder of the paper is organised as follows. Section 10.2 contains a brief review of the static model. Some basic elements of differential game theory are laid out in section 10.3. The setup of the dynamic model is described in section 10.4. Section 10.5 deals with the monopoly optimum, while section 10.6 describes the duopoly game. Section 10.7 contains some concluding remarks.

10.2 Preliminaries I: The static approach

Here I briefly summarise the static model. Consumers are uniformly distributed with density equal to one over the interval $[\Theta - 1\,, \Theta]$, with $\Theta > 1$. Therefore, the total population of consumers is represented by a unit square. Each consumer is indexed by a marginal willingness to pay for quality $\theta \in$

[4] The time dimension can play a decisive role in reversing the usual profit ranking between high- and low-quality suppliers also for other reasons, as shown in a differential game by Colombo and Lambertini (2003). Provided the low-quality firm is more efficient than the high-quality firm in terms of advertising activity or investment in productive capacity, then offering a superior quality does not necessarily entail higher profits than the rival's.

[5] For an overview, see Dockner *et al.* (2000). Another instance of the same kind is in Cellini and Lambertini (1998), where Cournot and Bertrand equilibria with differentiated products *à la* Singh and Vives (1984) are derived using the Ramsey capital accumulation model. However, it can also be shown that this correspondence does not hold when using the Solow model (Fudenberg and Tirole, 1983; Fershtman and Muller, 1984; Reynolds, 1987, 1991; Cellini and Lambertini, 2001).

$[\Theta - 1, \Theta]$, and his net utility from consumption is:

$$U = \begin{cases} \theta q_i - p_i \geq 0 \text{ if he buys} \\ 0 \text{ if he doesn't buy} \end{cases} \tag{10.1}$$

where p_i is the price of variety i. First, I focus upon the monopoly case and the social optimum.

10.2.1 Single-good monopolist

The cost of developing variety i is $C_i = cq_i^2$, with $c > 0$. The monopolist sells a single variety, and chooses price and quality so as to maximise profits:

$$\pi_M = px - cq^2 , \tag{10.2}$$

where $x = \Theta - p/q$ is market demand. From the first order conditions (FOCs):

$$\begin{aligned} \frac{\partial \pi_M}{\partial p} &= \Theta - \frac{2p}{q} = 0 \\ \frac{\partial \pi_M}{\partial q} &= \frac{p^2}{q^2} - 2cq = 0 \end{aligned} \tag{10.3}$$

one obtains $p_M^* = \Theta^3/(16c)$ and $q_M^* = \Theta^2/(8c)$. The associated equilibrium output is:

$$x_M^* = \begin{cases} \dfrac{\Theta}{2} < 1 \text{ for all } \Theta \in [1, 2) \\ 1 \text{ for all } \Theta \geq 2 \end{cases} \tag{10.4}$$

while profits and consumer surplus are:

$$\pi_M^* = \frac{\Theta^4}{64c}; \; CS_M^* = \int_{p/q}^{\Theta} (sq_M^* - p_M^*)\, d\theta = \frac{\Theta^4}{64c} = \pi_M^* . \tag{10.5}$$

10.2.2 Social optimum with a single good

Suppose the firm is managed by a benevolent planner maximising social welfare, $SW_{SP} = \pi + CS$, w.r.t. p and q. Subscript SP stands for *social planning*. The FOCs are:

$$\begin{aligned} \frac{\partial SW}{\partial p} &= \frac{p}{q} = 0 \\ \frac{\partial SW}{\partial q} &= \frac{\Theta^2}{2} + \frac{p^2}{2q^2} - 2cq = 0 \end{aligned} \tag{10.6}$$

yielding $p_{SP}^* = 0$ and $q_{SP}^* = \Theta^2/(4c)$. As we already know from Spence (1975) and Mussa and Rosen (1978), the comparison between q_{SP}^* and q_M^* entails the following result:

Proposition 10.1 *Under uniform consumer distribution and convex costs of quality improvement, the profit-seeking monopolist undersupplies quality as compared to the social optimum.*

It is also immediately verified that the planner serves all consumers in the market, irrespective of the level of Θ. Hence, for all $\Theta \in [1, 2)$, profit-seeking behaviour also causes a downward distortion in the output level.

If $c = 0$, i.e., quality improvements are costless, or $C_i = k$, i.e., R&D costs are exogenous, it is a straightforward task to prove that the partial derivative of both profits and social welfare w.r.t. quality is always positive. This makes it necessary to introduce an upper bound to the technologically feasible quality, say, Q. Then, one gets $q_{SP}^* = q_M^* = Q$, and there remains only the monopoly output restriction for all $\Theta \in [1, 2)$.

Under both regimes, the firm never finds it optimal to supply two varieties. The proof of this result is easy to obtain and it is omitted for brevity.

10.2.3 The duopoly game

Leaving unchanged the assumptions on consumer preferences, I can proceed with the outline of the two-stage model analysed in several contributions (Choi and Shin, 1992; Dutta *et al.*, 1995; Lambertini *et al.*, 2002, *inter alia*). Two single-product firms, labelled as H and L, supply goods of qualities $Q > q_H \geq q_L \geq 0$. All variable costs are assumed to be nil, and quality improvements involve an exogenous R&D effort k_i. The unit cost of capital is ρ, and therefore the R&D cost is ρk_i. As this cost is not explicitly defined as a function of quality, it does not affect first order conditions. It is sensible to assume that a higher quality requires a higher R&D effort, and therefore $k_H > k_L$.[6]

Demands for the two goods are:

$$x_H = \Theta - \theta_H \; ; \; x_L = \theta_H - \theta_L \; , \tag{10.7}$$

where θ_H is the marginal willingness to pay for quality characterising the consumer who is indifferent between q_H and q_L at the price vector $\{p_H \, , \, p_L\}$, i.e., it is the solution to:

$$\theta_H q_H - p_H = \theta_H q_L - p_L \Leftrightarrow \theta_H = \frac{p_H - p_L}{q_H - q_L} \, , \tag{10.8}$$

[6] Alternatively, one could assume that the development cost borne by firm i is $C_i = cq_i^2$. See Motta (1993) and ch. 8. The reason for assuming an exogenous R&D cost will become clear in the remainder.

while θ_L is the marginal evaluation of quality associated with the consumer who is indifferent between buying the low quality good and not buying at all, $\theta_L = p_L/q_L$. Firms' profits, which coincide with revenues, are:

$$\pi_H = p_H\left(\Theta - \frac{p_H - p_L}{q_H - q_L}\right) - \rho k_H \; ; \; \pi_L = p_L\left(\frac{p_H - p_L}{q_H - q_L} - \frac{p_L}{q_L}\right) - \rho k_L \, . \quad (10.9)$$

Firms play simultaneously a non-cooperative two-stage game, where they set qualities in the first stage and price in the second. As usual, the solution concept is subgame perfection by backward induction. The outcome is summarised in the following Proposition, the complete proof of which can be found in Choi and Shin (1992) (superscript *sp* stands for *subgame perfect*):

Proposition 10.2 *At the subgame perfect equilibrium,*

- *qualities are* $q_H^{sp} = Q$ *and* $q_L^{sp} = 4Q/7$;
- *output levels are* $x_H^{sp} = 7\Theta/12$ *and* $x_L^{sp} = 7\Theta/24 = x_H^{sp}/2$;
- *prices are* $p_H^{sp} = \Theta Q/4$ *and* $p_L^{sp} = \Theta Q/14$;
- *profits are* $\pi_H^{sp} = \dfrac{7\Theta^2 q_H}{48} - \rho k_H \; ; \; \pi_L^{sp} = \dfrac{\Theta^2 q_H}{48} - \rho k_L$.

The result that $q_L^{sp} = 4Q/7$ tells that the high quality is exactly equal to the highest (and exogenously given) feasible quality Q, while the low quality locates slightly above the middle of the quality spectrum. In the remainder of the paper, I will refer to this result as the *4/7 rule.* The above Proposition also allows to establish that $x_H^{sp} + x_L^{sp} < 1$ for all $\Theta < 8/7$. Hence, for all $\Theta \geq 8/7$, the demand functions (10.7) are not valid and the model must be re-specified with $x_L = \theta_H - (\Theta - 1)$.[7]

Now observe that

Corollary 10.1 *The duopoly equilibrium is sustainable iff:*

1] $\pi_i^{sp} > 0$ *for all* i, *which entails* $q_H > \max\left\{\dfrac{48\rho k_L}{\Theta^2}, \dfrac{48\rho k_H}{7\Theta^2}\right\}$, *where*

$$\max\left\{\frac{48\rho k_L}{\Theta^2}, \frac{48\rho k_H}{7\Theta^2}\right\} = \begin{cases} \dfrac{48\rho k_H}{7\Theta^2} & \text{for all } k_H > 7k_L \\ \dfrac{48\rho k_L}{\Theta^2} & \text{for all } k_H \in (k_L\,, 7k_L) \end{cases}$$

and

[7] See Tirole (1988, appendix to ch. 7) and Rosenkranz (1995).

2] $\pi_H^{sp} > \pi_L^{sp}$, *which requires* $k_H - k_L < \dfrac{\Theta^2 q_H}{8\rho}$.

In particular, if the second condition in corollary 10.1 is not met, there is no incentive for either firm to enter first and offer the high-quality good. Consequently the market remains inactive because producing the superior variety is not convenient.

Moreover, $x_H^{sp} = 7\Theta/12$ implies that $x_L^{sp} > 0$ iff $\Theta < 12/7$. Therefore, Proposition 1 produces the following corollary:

Corollary 10.2 *For all* $\Theta \geq 12/7$, *the market is monopolised by the high quality firm.*

Corollary 10.2 is an instance of the so-called *finiteness property* (Shaked and Sutton, 1983), which establishes that the demand structure of a vertically differentiated market allows for a finite number of firms operating with positive demand and profits at the subgame perfect equilibrium. In particular, the above case is what Shaked and Sutton label as a *natural monopoly*. They use consumer income, while here I use the marginal evaluation for quality, as in Mussa and Rosen (1978) and Gabszewicz and Thisse (1979, 1980). It can be easily shown that the two approaches are equivalent, provided that consumer's utility function is concave in income.[8]

The finiteness property can be shown to hold also in models where quality affects fixed costs (see, e.g., Motta, 1993; Lehmann-Grube, 1997; and Lambertini, 1999). Of course, considering any endogenous cost function defined in terms of quality entails that the *4/7 rule* does not hold in slightly more sophisticated reformulations of the static model.

For future reference, it is worth noting that the (exogenously imposed) upper bound of the quality spectrum, Q (the highest technologically feasible quality), generates the corner solution $q_H^{sp} = Q$, as the revenues of firm H are everywhere increasing in q_H. Being quality improvements costless, without such a boundary q_H would become infinitely high. Moreover, it is reasonable to think that the upper bound Q should be endogenised, in the sense that firms' R&D investments (in particular, firm H's) should increase the feasible quality range.[9]

[8] Under this condition, $\theta = \beta/u_y$, where $u_y \equiv \partial u(y)/\partial y$ is the marginal utility of income and β is a positive parameter. If $u_{yy} \equiv \partial^2 u(y)/\partial y^2 \leq 0$, the marginal willingness to pay for quality increases as income increases (see Tirole, 1988, ch. 2).

[9] Of course this idea can be developed in a static context by assuming, e.g., a convex cost of quality improvement. This is investigated in Motta (1993), Lambertini (1996, 1999, 2000), and Lehmann-Grube (1997), *inter alia*. However, in such a case the derivation of the subgame perfect equilibrium requires numerical calculations.

10.3 Preliminaries II: Elements of differential game theory

Consider a game to be played over continuous time, $t \in [0, \infty)$.[10] Define the set of players as $\mathbb{P} \equiv \{1, 2, 3, ...N\}$. Moreover, let $y_i(t)$ and $z_i(t)$ define, respectively, the state variable and the control variable pertaining to player i. For simplicity, we consider the case where only one state and one control are associated to every single player. The dynamics of player i's state variable is described by the following:

$$\frac{dy_i(t)}{dt} \equiv \dot{y}_i(t) = f_i\left(\{y_i(t)\}_{i=1}^N, \{z_i(t)\}_{i=1}^N\right) \tag{10.10}$$

where $\{y_i(t)\}_{i=1}^N \equiv Y(t)$ is the vector of state variables at time t, and $\{z_i(t)\}_{i=1}^N$ is the vector of players' actions at the same date, i.e., it is the vector of the values of control variables at time t. That is, in the most general case, the dynamics of the state variable associated to player i depends on all state and control variables associated to all players involved in the game. The value of the state variables at $t = 0$ is assumed to be known: $\{y_i(0)\}_{i=1}^N = \{y_{0,i}\}_{i=1}^N$.

Each player has an objective function, defined as the discounted value of the flow of payoffs over time. The instantaneous payoff depends upon the choices made by player i as well as its rivals, that is:

$$\pi_i(t) = \pi_i\left(y_i(t), Y_{-i}(t), z_i(t), Z_{-i}(t)\right), \tag{10.11}$$

where $Y_{-i}(t)$ is the vector of the values of states of all other players, at time t, and $Z_{-i}(t)$ summarises the actions of all other players at time t. Player i's objective is then

$$\max_{u_i(t)} J_i \equiv \int_0^\infty \pi_i(., t)e^{-\rho t}dt \tag{10.12}$$

subject to the dynamic constraint represented by the behaviour of the state variables (10.10) for $i = 1, ...N$, and given the initial conditions on states, which are assumed to be known to all players. The factor $e^{-\rho t}$ discounts future gains, and the discount rate ρ is assumed to be constant and common to all players. In order to solve his optimisation problem, each player defines a strategy.

We will consider two different information structures, to which two different equilibrium concepts are associated:

[10] The game can be reformulated in discrete time without significantly affecting its qualitative properties. For further details, see Başar and Olsder (1982, 1995[2]).

Definition 10.1: Open-loop information *In this case, at any time t each player's information set includes only calendar time, but not the history and the current stocks of states. Hence, the optimal choice of control z_i is conditional on current time only, so that the optimal open-loop strategy appears as $z_i^* = z_i^*(t)$.*

Definition 10.2: Closed-loop memoryless information *In this case, at any time t each player's information includes calendar time and the state vector $Y(t)$. Hence, the optimal closed-loop memoryless strategy is $z_i^* = z_i^*(t, Y(t))$.*

By 'memoryless', it is meant that the game history in itself is not relevant for the choice of optimal behaviour at time t, only the consequences of the history are important, as they are reflected in the current state vector. However, taking into account the current state vector allows each player to optimally react to the other players' behaviour, because of the effect exerted by the other players' strategies through the vector $Y(t)$. This entails that the closed-loop memoryless equilibrium is subgame perfect. On the contrary, this is not true in general for the open-loop Nash equilibrium, because it consists of strategies which are only a function of time and do not properly account for strategic interaction at intermediate times between the initial date and the steady state.

In the literature on differential games, one usually refers to the concepts of *weak* and *strong* time consistency. Essentially, for a solution to be time consistent (in either sense), the players should have no rational reason, at any future date during the game, to deviate from the equilibrium strategies. The difference between these two properties can be outlined as follows:

Definition 10.3: Weak time consistency *Consider a game played over $t = [0, \infty)$ and examine the trajectories of the state variables generated under the Nash equilibrium, denoted by $y(t)$. The equilibrium is weakly time consistent if its truncated part in the time interval $t = [T, \infty)$, with $T \in (0, \infty)$, represents an equilibrium also for any subgame starting from $t = T$, and from the initial condition $y_T = y(t = T)$.*

Definition 10.4: Strong time consistency *Consider a game played over $t = [0, \infty)$ and examine the trajectories of the state variables generated under the Nash equilibrium, denoted by $x(t)$. The equilibrium is strongly time consistent, if its truncated part is an equilibrium for the subgame, independently of the conditions regarding state variables y at time T, $\{y_i(T)\}_{i=1}^{N}$.*

Strong time consistency requires the ability on the part of each player to account for the rival's behaviour at any point in time, i.e., it is, in general, an attribute of closed-loop equilibria, and corresponds to subgame perfectness. Weak time consistency is a milder requirement and does not ensure, in general, that the resulting Nash equilibrium be subgame perfect.[11]

In the particular case where the feedback exerted by the other players' strategies on player i's optimal choice at any t during the game is nil, the resulting optimal closed-loop strategy is $z_i^*(t)$, i.e., it is no longer a function of $Y(t)$. As a result, the closed-loop memoryless strategy collapses into the open-loop one, and the open-loop equilibrium outcome $\{z_i^*(t)\}_{i=1}^N$ is strongly time consistent, or subgame perfect. In the remainder of the section, we clarify under what circumstances this is going to happen.

10.3.1 Solution methods

Under the open-loop solution concept, the Hamiltonian of player i writes as follows:

$$\mathcal{H}_i \equiv e^{-\rho t}\left[\pi_i\left(y_i(t), Y_{-i}(t), z_i(t), Z_{-i}(t)\right) + \lambda_{ii}(t) \cdot f_i\left(\{y_i(t)\}_{i=1}^N, \{z_i(t)\}_{i=1}^N\right)\right.$$

$$\left. + \sum_{j \neq i} \lambda_{ij}(t) \cdot f_j\left(\{y_i(t)\}_{i=1}^N, \{z_i(t)\}_{i=1}^N\right)\right], \qquad (10.13)$$

where $\lambda_{ij}(t) = \mu_{ij}(t)e^{\rho t}$ is the co-state variable (evaluated at time t) associated with the state variable $y_j(t)$. If the evolution of the state variable $y_i(t)$ depends only upon $\{y_i(t), z_i(t)\}$, i.e., it is independent of $Z_{-i}(t)$ and $y_{-i}(t)$ and (10.10) simplifies as $\dot{y}_i(t) = f_i(y_i(t), z_i(t))$, then one can set $\lambda_{ij}(t) = 0$ for all $j \neq i$, which entails that the Hamiltonian of player i can be written by taking into account the dynamics of i's state variable only.

The first order condition on the control variable $u_i(t)$ is:

$$\frac{\partial \mathcal{H}_i(., t)}{\partial z_i(t)} = 0 \qquad (10.14)$$

and the adjoint equations concerning the dynamics of state and co-state variables are as follows:

$$-\frac{\partial \mathcal{H}_i(., t)}{\partial y_j(t)} = \frac{\partial \lambda_{ij}(t)}{\partial t} - \rho\lambda_{ij}, \forall j = 1, 2...N \qquad (10.15)$$

[11] For a more detailed analysis of these issues, see Dockner *et al.* (2000, Section 4.3, pp. 98-107); see also Başar and Olsder (1982, 1995[2], ch. 6).

They have to be considered along with the initial conditions $\{y_i(0)\}_{i=1}^{N} = \{y_{i0}\}_{i=1}^{N}$ and the transversality conditions, which set the final value (at $t = \infty$) of the state and/or co-state variables. In problems defined over an infinite time horizon, it is usually set:

$$\lim_{t\to\infty} \lambda_{ij}(t) \cdot y_j(t) = 0, \forall j = 1, 2...N. \tag{10.16}$$

From (10.14) one obtains the instantaneous best reply of player i, which can be differentiated with respect to time to yield the kinematic equation of the control variable $z_i(t)$. This, in combination with the state equations (10.10) and the adjoint equations (10.15) permits us to identify the open-loop Nash equilibrium (or equilibria) of the game; the optimal time path of the control variables will depend on t only. Moreover, the simultaneous consideration of the dynamic constraint leads to a dynamic system, whose properties can be easily studied. One is generally interested in proving the existence and characterising the features of steady state(s), and in studying its (their) stability properties.

Under the closed-loop memoryless information structure, the Hamiltonian of player i is the same as in (10.13). The relevant first order conditions and the adjoint equations are:

$$\frac{\partial \mathcal{H}_i(.,t)}{\partial z_i(t)} = 0 ; \tag{10.17}$$

$$-\frac{\partial \mathcal{H}_i(.,t)}{\partial y_j(t)} - \sum_{h\neq j} \frac{\partial \mathcal{H}_i(.,t)}{\partial z_h(t)} \frac{\partial z_h^*(t)}{\partial y_j(t)} = \frac{\partial \lambda_{ij}(t)}{\partial t} - \rho\lambda_{ij} , \forall j = 1, 2...N; \tag{10.18}$$

along with the initial conditions $\{y_i(0)\}_{i=1}^{N} = \{y_{0,i}\}_{i=1}^{N}$ and the transversality conditions

$$\lim_{t\to\infty} \lambda_{ij}(t) \cdot y_j(t) = 0 , \forall j = 1, 2...N. \tag{10.19}$$

The terms

$$\frac{\partial \mathcal{H}_i(.,t)}{\partial z_h(t)} \frac{\partial z_h^*(t)}{\partial y_j(t)} \tag{10.20}$$

appearing in the adjoint equations capture the strategic interaction through the feedback from states to controls, which is by definition absent under the open-loop solution concept. In equations (10.18) and (10.20), $z_h^*(t)$ is the solution to the first order condition of firm h w.r.t. her control variable. Whenever the expression in (10.20) is zero for all h, then the closed-loop memoryless equilibrium collapses into the open-loop Nash equilibrium, in the sense that the time path of all relevant variables under the two different information structures coincide. This can happen either because:

$$\frac{\partial \mathcal{H}_i(.,t)}{\partial z_h(t)} = 0 \text{ for all } h \neq i , \tag{10.21}$$

which obtains if the Hamiltonian of player i is a function of his control variable but not of the rivals'; or because:

$$\frac{\partial z_h^*(t)}{\partial y_j(t)} = 0 \text{ for all } h \neq j\,, \tag{10.22}$$

which means that the first order condition of a player with respect to his control variable does not contain the state variables pertaining to any other players. Of course, it could also be that (10.21) and (10.22) hold simultaneously. In such cases, the open-loop Nash equilibrium qualifies as a strongly time consistent (subgame perfect) solution of the game.

10.3.2 Elements of stability analysis

In analysing dynamic settings (not only differential games but also optimization problems with a single agent), one is also interested in evaluating whether a *steady state* exists. By this, we mean a vector of variables which possesses the desirable property that, whenever players reach the steady state, then all the relevant variables remain constant thereafter.

A steady state equilibrium may not exist; if it does exist, still it may not be unique. Moreover, a steady state equilibrium may exhibit different features as far as its stability properties are concerned. In particular, the following taxonomy applies. A steady state can be, alternatively:

A. a *stable (unstable) node.* This holds when the system non-cyclically converges to (diverges from) that steady state, regardless of where it starts from;

B. *stable along a saddle path.* This holds when there exists one and only one time path leading to the steady state;

C. a *stable (unstable) focus.* This holds when the system cyclically converges to (diverges from) the steady state;

D. a *vortex.* This holds when the system perpetually orbits around the steady state.

For illustrative purposes, I confine my attention here to the case where the game enjoys symmetry properties such that the stability analysis can be carried out as if the setting could be treated as a single-agent optimization problem.[12] Define the steady state as the vector $\{y^*, z^*\}$. This vector is the

[12] This is the case if players' maximands are fully symmetric, so that the state and control variables are symmetric in equilibrium.

outcome of the dynamic system:

$$\begin{cases} \dfrac{dy(t)}{dt} \equiv \dot{y}(t) = f(y,z) = 0 \\ \dfrac{dz(t)}{dt} \equiv \dot{z}(t) = g(y,z) = 0 \end{cases} \tag{10.23}$$

The dynamic equations in (10.23) can be linearised around the steady state point through a first order Taylor expansion, so that the system (10.23) can be written in matrix notation as follows:

$$\begin{bmatrix} \dot{y} \\ \\ \dot{z} \end{bmatrix} = \Xi \begin{bmatrix} (y - y^*) \\ \\ (z - z^*) \end{bmatrix} + \Psi \tag{10.24}$$

where Ξ is the Jacobian matrix of partial derivatives evaluated at $\{y^*, z^*\}$:

$$\Xi = \begin{bmatrix} f_y & f_z \\ g_y & g_z \end{bmatrix} \Bigg|_{y^*,z^*}$$

and $\Psi = \{f(y^*, z^*),\ g(y^*, z^*)\}$ is a column vector whose components are zero, since $f(\cdot,\cdot)$ and $g(\cdot,\cdot)$ are nil when evaluated at $\{y^*, z^*\}$. The stability properties of the system in the neighbourhood of the steady state depend upon the trace and determinant of matrix Ξ. In particular, the system produces a saddle when the determinant is negative. Of course, in looking for steady states, we have to ascertain whether optimality conditions (10.17-10.18) are indeed compatible with $dy(t)/dt = 0$ and $dz(t)/dt = 0$.

As a last remark on steady state Nash equilibria, observe that the analysis of the properties of a dynamic system is conceptually distinct from and technically independent of the issue of the equilibrium of a differential game. We have a Nash equilibrium when each agent plays the best response to all his opponents' actions. From the standpoint of the analysis of a dynamic system, "equilibrium" means that variables are stationary over time. Both issues are relevant when we focus upon a steady state Nash equilibrium, i.e., a state where the system (the market, if we refer to the theory of industrial organization) stays forever after, provided each agent plays his optimal strategy.

Needless to say, the optimal control problem with a single agent obtains as a special case of the differential game, by setting $N = 1$.

10.4 The dynamic model

The market exists over $t \in [0, \infty)$. At any t, as in the static version, a constant population of consumers is uniformly distributed with density equal to one over the interval $[\Theta - 1, \Theta]$, with $\Theta \geq 1$. Therefore, the total mass of consumers is also equal to 1. The consumers who are able to buy do so at each instant $t \in [0, \infty)$. To this regard, it is worth stressing that, as purchases repeat over time, this model suits the case of non-durable goods. Each consumer is characterised by a marginal willingness to pay for quality $\theta \in [\Theta - 1, \Theta]$, and his net instantaneous utility from consumption is now defined as:

$$U = \begin{cases} \theta q_i(t) - p_i(t) \geq 0 \text{ if he buys} \\ 0 \text{ if he doesn't buy} \end{cases} \tag{10.25}$$

where $q_i(t)$ is the quality and $p_i(t)$ is the price of good i at time t. The quality of variety i increases over time according to the following dynamics:

$$\frac{dq_i(t)}{dt} \equiv \dot{q}_i = ak_i(t) - \delta q_i(t) , \tag{10.26}$$

where $a > 0$ and $\delta \in [0, 1]$ is a constant depreciation (or obsolescence) rate affecting quality. The initial condition for good i is $q_i(0) = q_{i0} \geq 0$. The instantaneous cost associated to the R&D activity is $C_i(k_i(t)) = b[k_i(t)]^2$, $b > 0$. This amounts to saying that decreasing returns in the production of quality take place through instantaneous R&D costs. There are no costs other than $C_i(k_i(t))$. That is, operative production costs are assumed to be nil, and therefore instantaneous profits generated by variety i are given by the difference between revenues and the cost of investment.

10.5 Monopoly

The first investigation of dynamic monopoly dates back to Evans (1924) and Tintner (1937), who analysed the pricing behaviour of a firm subject to a U-shaped variable cost curve.[13] The analysis of intertemporal capital accumulation appeared later on (see Eisner and Strotz, 1963, *inter alia*).

Another dynamic tool which has received a considerable amount of attention is advertising, ever since Vidale and Wolfe (1957) and Nerlove and Arrow (1962).[14]

[13] See Chiang (1992) for a recent exposition of the original model by Evans, as well as later developments.

[14] For exhaustive surveys, see Sethi (1977); Jørgensen (1982); Feichtinger and Jørgensen (1983); Erickson (1991); Feichtinger, Hartl and Sethi (1994); Dockner, Jørgensen, Van

10.5.1 Single-product monopoly

The market is supplied by a profit-seeking monopolist. The demand function at any time t is:

$$x(t) = \Theta - \overline{\theta}(t) = \Theta - \frac{p(t)}{q(t)} \tag{10.27}$$

where subscript M stands for *monopoly*, and $\overline{\theta}(t) \equiv p(t)/q(t)$ is the marginal willingness to pay of the marginal consumer. The instantaneous monopoly profits are:

$$\pi_M(t) = p(t)\left(\Theta - \frac{p(t)}{q(t)}\right) - b\left[k(t)\right]^2 . \tag{10.28}$$

The monopolist's problem consists in maximizing w.r.t. $p(t)$ and $k(t)$ the following Hamiltonian function:

$$\mathcal{H}_M(t) = e^{-\rho t}\left\{p(t)\left(\Theta - \frac{p(t)}{q(t)}\right) - b\left[k(t)\right]^2 + \lambda_M(t)\left[ak(t) - \delta q(t)\right]\right\} , \tag{10.29}$$

where $\lambda_M(t) = \mu_M(t)e^{\rho t}$, and $\mu_M(t)$ is the co-state variable associated to $q(t)$. The initial condition is $q(0) = q_0 \geq 0$. The outcome is summarised by the following Proposition:

Proposition 10.3 *The monopolist's optimal R&D effort and quality are:*

$$k_M^* = \frac{a\Theta^2}{8b\left(\rho + \delta\right)} , \; q_M^* = \frac{ak_M^*}{\delta} = \frac{a^2\Theta^2}{8b\delta\left(\rho + \delta\right)} .$$

At $\{k_M^* , q_M^*\}$, partial (respectively, full) market coverage obtains for all $\Theta \in [1, 2)$ (respectively, $\Theta \geq 2$). For all $\delta \in (0, 1]$, the pair $\{k_M^* , q_M^*\}$ is a saddle point.

Proof. The necessary conditions for a path to be optimal are:[15]

$$\frac{\partial \mathcal{H}_M}{\partial p} = \Theta - \frac{2p}{q} = 0 ; \tag{10.30}$$

$$\frac{\partial \mathcal{H}_M}{\partial k} = -2bk + \lambda_M a = 0 ; \tag{10.31}$$

Long and Sorger (2000). For duopoly models with dynamic pricing and advertising, see in particular Leitmann and Schmitendorf (1978), and Feichtinger (1983).

[15]Second order conditions for concavity are also met. They are omitted for the sake of brevity. Likewise, I will omit, in this proof as well as in the next ones, the indication of time and exponential discounting.

$$-\frac{\partial \mathcal{H}_M}{\partial q} = \frac{\partial \mu_M}{\partial t} \Rightarrow \frac{\partial \lambda_M}{\partial t} = (\rho + \delta) \lambda_M - \left(\frac{p}{q}\right)^2 ; \tag{10.32}$$

$$\lim_{t \to \infty} \mu_M \cdot q = 0 . \tag{10.33}$$

From (10.30), I immediately obtain:

$$p_M = \frac{\Theta q_M}{2} . \tag{10.34}$$

Incidentally, note that the optimum price choice does not require a dynamic analysis. This is the reason why the Proposition refers to a steady state defined w.r.t. R&D effort and quality.

Then, from (10.31), I obtain:

$$\lambda_M = \frac{2bk}{a} \Leftrightarrow k = \frac{a\lambda_M}{2b} \tag{10.35}$$

from which I can derive the dynamics of k :

$$\frac{dk}{dt} = \frac{a}{2b} \cdot \frac{\partial \lambda_M}{\partial t} . \tag{10.36}$$

Using (10.32), (10.34) and (10.35), the above expression can be conveniently rewritten as follows:

$$\frac{dk}{dt} = k(\rho + \delta) - \frac{a\Theta^2}{8b} , \tag{10.37}$$

which is equal to zero at $k_M^* = a\Theta^2 / [8b(\rho + \delta)]$. This solution can be plugged into $dq/dt = 0$ to obtain $q_M^* = a^2\Theta^2 / [8b\delta(\rho + \delta)]$. In such a steady state, we have

$$p_M^* = \frac{\Theta q_M^*}{2} \ ; \ x_M^* = \begin{cases} \dfrac{\Theta}{2} < 1 \text{ for all } \Theta \in [1, 2) \\ 1 \text{ for all } \Theta \geq 2 \end{cases} \tag{10.38}$$

If $\Theta \in [1, 2)$, the steady state equilibrium profits are:

$$\pi_M^* = \frac{a^2\Theta^4 (2\rho + \delta)}{64b\delta(\rho + \delta)^2} , \tag{10.39}$$

while consumer surplus is:

$$CS_M^* = \int_{\overline{\theta}}^{\Theta} (\theta q_M^* - p_M^*) \, d\theta = \frac{a^2\Theta^4}{64b\delta(\rho + \delta)} . \tag{10.40}$$

Therefore, social welfare is:

$$SW_M^* = \frac{a^2\Theta^4 (3\rho + 2\delta)}{64b\delta(\rho + \delta)^2} . \tag{10.41}$$

Now, using (10.26) and (10.37), I can calculate the following:

$$\frac{\partial \dot{q}}{\partial q} = -\delta \, ; \, \frac{\partial \dot{q}}{\partial k} = a \, ; \tag{10.42}$$

$$\frac{\partial \dot{k}}{\partial q} = 0 \, ; \, \frac{\partial \dot{k}}{\partial k} = \rho + \delta \, ; \tag{10.43}$$

where:

$$\dot{q} \equiv \frac{dq}{dt} \, ; \, \dot{k} \equiv \frac{dk}{dt} \, . \tag{10.44}$$

The partial derivatives in (10.42-10.43) are the elements of the Jacobian matrix of the dynamic problem under examination. The trace and determinant are, respectively:

$$Tr\,(\Xi) = \rho \, ; \, \Delta\,(\Xi) = -\delta\,(\rho + \delta) < 0 \text{ for all } \delta \in (0, 1] \, .$$

This shows that the pair $\{k_M^*, q_M^*\}$ is a saddle point. As a final remark, it is worth stressing that, if δ were nil, the system would produce no steady state. Accordingly, the monopolist would keep increasing the quality level forever. ■

10.5.2 The social optimum with a single variety

If a benevolent planner runs the firm, he aims at maximising the discounted value of social welfare defined as the sum of profits and consumer surplus:

$$\int_0^\infty SW(t) e^{\rho t} dt = \int_0^\infty \left[\pi(t) + CS(t) \right] e^{\rho t} dt \tag{10.45}$$

under the constraint (10.26). Instantaneous consumer surplus is given by:

$$CS(t) = \int_{\overline{\theta}}^{\Theta} \left[\theta q(t) - p(t) \right] d\theta \, . \tag{10.46}$$

Again, I carry out a partial equilibrium analysis, assuming that the planner may raise funds from other sources if any loss is caused by marginal cost pricing. The relevant Hamiltonian under social planning (SP) is:

$$\mathcal{H}_{SP} = e^{-\rho t} \left\{ p \left(\Theta - \frac{p}{q} \right) + \frac{\left[\Theta q - p \right]^2}{2q} - b \left[k \right]^2 + \lambda_{SP} \left[ak - \delta q \right] \right\} , \tag{10.47}$$

which has to be maximised w.r.t. p and k. The initial condition is the same as above. The socially efficient behaviour of the planner is summarised by the following:

Proposition 10.4 *The planner's optimal R&D effort and quality are:*

$$k_{SP}^* = \frac{a\Theta^2}{4b(\rho+\delta)}\,,\; q_{SP}^* = \frac{ak_{SP}^*}{\delta} = \frac{a^2\Theta^2}{4b\delta(\rho+\delta)}\,.$$

At $\{k_{SP}^*, q_{SP}^*\}$, full market coverage obtains for all Θ, due to marginal cost pricing. For all $\delta \in (0,1]$, the pair $\{k_{SP}^*, q_{SP}^*\}$ is a saddle point.

Proof. The necessary and sufficient conditions for a path to be optimal are:

$$\frac{\partial \mathcal{H}_{SP}}{\partial p} = -\frac{p}{q} = 0\,; \tag{10.48}$$

$$\frac{\partial \mathcal{H}_{SP}}{\partial k} = -2bk + \lambda_{SP}a = 0\,; \tag{10.49}$$

$$-\frac{\partial \mathcal{H}_{SP}}{\partial q} = \frac{\partial \mu_{SP}}{\partial t} \Rightarrow \frac{\partial \lambda_{SP}}{\partial t} = (\rho+\delta)\lambda_{SP} - \frac{\Theta^2}{2} - \frac{1}{2}\left(\frac{p}{q}\right)^2\,; \tag{10.50}$$

$$\lim_{t\to\infty} \mu_{SP}\cdot q = 0\,. \tag{10.51}$$

From (10.48), $p_{SP}^* = 0$, and therefore $x_{SP}^* = 1$. Then, from (10.49), we obtain the same as in (10.35); accordingly, capital evolves over time as in (10.36). This is an interesting point, as it establishes that, judging from the first order conditions on the R&D effort for the monopolist and the social planner, one cannot immediately draw the implication that the monopolist will distort the levels of investment (and therefore the resulting quality) as compared to the social optimum. As illustrated above, this is the case in the static setting. Here, as we are about to see, the distortion induced by the profit-seeking behaviour comes from the co-state equation, describing the dynamics of both λ_{SP} and k.

Using $p_{SP}^* = 0$ and (10.50), we have the following dynamics of the R&D investment:

$$\frac{dk(t)}{dt} = k(\rho+\delta) - \frac{a\Theta^2}{4b}\,, \tag{10.52}$$

which is zero at

$$k_{SP}^* = \frac{a\Theta^2}{4\rho^4}\,. \tag{10.53}$$

This value can be plugged into $dq/dt = 0$ to obtain $q_{SP}^* = a^2\Theta^2/[4b\delta(\rho+\delta)]$. Consumer surplus and social welfare at equilibrium are:

$$CS_{SP}^* = \frac{a^2\Theta^4}{8b\delta(\rho+\delta)}\,;\; SW_{SP}^* = \frac{a^2\Theta^4(2\rho+\delta)}{16b\delta(\rho+\delta)^2}\,. \tag{10.54}$$

The analysis of the Jacobian of this problem is largely analogous to that carried out for the monopoly case, and therefore it is omitted. ■

A straightforward implication of Propositions 10.3-10.4 is:

Corollary 10.3 *In the social optimum, R&D investment and product quality are higher than in the monopoly optimum. Output is larger under social planning than under profit-seeking monopoly for all* $\Theta < 2$.

10.5.3 The two-product monopolist

Now suppose the monopolist develops two varieties, $\infty > q_H(t) \geq q_L(t) \geq 0$. Each quality entails an instantaneous R&D cost $C_i(k_i(t)) = b[k_i(t)]^2$, and increases over time according to (10.26). Instantaneous market demands are:

$$x_H(t) = \Theta - \theta_H(t)\ ;\ x_L(t) = \theta_H(t) - \theta_L(t)\ , \tag{10.55}$$

where $\theta_H(t)$ is the marginal willingness to pay for quality characterising the consumer who is indifferent between $q_H(t)$ and $q_L(t)$ at the price vector $\{p_H(t), p_L(t)\} : \theta_H(t) = [p_H(t) - p_L(t)] / [q_H(t) - q_L(t)]$, while $\theta_L(t) = p_L(t)/q_L(t)$. The monopolist aims at maximising w.r.t. prices and investment levels the discounted flow of profits:

$$J_M \equiv \int_0^\infty \pi_M(t)\, e^{-\rho t} dt \tag{10.56}$$

where:

$$\begin{aligned}\pi_M(t) \;=\;& p_H(t)\left(\Theta - \frac{p_H(t) - p_L(t)}{q_H(t) - q_L(t)}\right) + p_L(t)\left(\frac{p_H(t) - p_L(t)}{q_H(t) - q_L(t)} - \frac{p_L(t)}{q_L(t)}\right) + \\ & -b[k_H(t)]^2 - b[k_L(t)]^2, \end{aligned}\tag{10.57}$$

under the constraints given by co-state equations (10.26). Writing the Hamiltonian and taking the FOCs w.r.t. $p_H(t)$ and $p_L(t)$, one can obtain the optimal prices:

$$p_i^* = \frac{\Theta q_i(t)}{2}, \tag{10.58}$$

which can be plugged into the demand functions, to yield:

$$x_H^* = \frac{\Theta}{2}\ ;\ x_L^* = 0. \tag{10.59}$$

Therefore, the following can be stated:

Proposition 10.5 *The profit-seeking monopolist supplies a single quality.*

This result closely recalls the analogous conclusion holding in the static model. Likewise, the optimum problem of a welfare-maximising social planner can be solved to verify that the same property applies.

10.5.4 Social planning with two varieties

With two varieties, the consumer surplus at any time t is defined as follows:

$$CS(t) = \int_{\theta_H(t)}^{\Theta} [\theta q_H(t) - p_H(t)]\, d\theta + \int_{\theta_L(t)}^{\theta_H(t)} [\theta q_L(t) - p_L(t)]\, d\theta\,. \quad (10.60)$$

Together with profits (10.57), (10.60) defines instantaneous social surplus $SW_{SP}(t)$. The planner must choose $k_i(t)$ and $p_i(t)$ so as to maximise the discounted social welfare:

$$J_M \equiv \int_0^{\infty} SW_{SP}(t)\, e^{-\rho t} dt \quad (10.61)$$

under the dynamic constraints (10.26). The FOCs w.r.t. prices are:

$$\frac{\partial \mathcal{H}_{SP}}{\partial p_H} = -\frac{p_H - p_L}{q_H - q_L} = 0\,; \quad (10.62)$$

$$\frac{\partial \mathcal{H}_{SP}}{\partial p_L} = -\frac{p_L q_H - p_H q_L}{(q_H - q_L)\, q_L} = 0\,. \quad (10.63)$$

The first implies $p_H = p_L$. This, in turn, entails that the second is met only in $q_L = q_H$. Therefore:

Proposition 10.6 *Like the profit-seeking monopolist, also the social planner supplies a single quality.*

10.6 The duopoly game

Two single-product firms, labelled as H and L, supply goods of qualities $\infty > q_H(t) \geq q_L(t) \geq 0$. The discount rate $\rho > 0$ is common to both firms. The co-state equations are defined as:

$$\frac{dq_H(t)}{dt} = ak_H(t) - \delta q_H(t)\,; \quad (10.64)$$

$$\frac{dq_L(t)}{dt} = ak_L(t)\,. \quad (10.65)$$

Note that no depreciation appears in $dq_L(t)/dt$. An economic explanation can be that depreciation only affects the top feasible quality supplied in the market, which in this case is endogenously determined by $q_H(t)$, because richer customers are more demanding than others, and therefore they ask for more. This translates into the presence of δ in $dq_H(t)/dt$. Instead, those who purchase $q_L(t)$ are just happy with it.[16] The initial condition for firm i is $q_i(0) = q_{i0} \geq 0$. The instantaneous cost associated to the R&D activity is $C_i(k_i(t)) = b[k_i(t)]^2$, and each firm bears no costs other than $C_i(k_i(t))$.

The definition of market demands is analogous to the static setup and the two-product monopoly. At any t, market demands for the two varieties are defined as in (10.55). Accordingly, instantaneous profits are:

$$\pi_H(t) = p_H(t)\left(\Theta - \frac{p_H(t) - p_L(t)}{q_H(t) - q_L(t)}\right) - b[k_H(t)]^2 ; \qquad (10.66)$$

$$\pi_L(t) = p_L(t)\left(\frac{p_H(t) - p_L(t)}{q_H(t) - q_L(t)} - \frac{p_L(t)}{q_L(t)}\right) - b[k_L(t)]^2 , \qquad (10.67)$$

provided that $x_H(t) + x_L(t) \leq 1$.

Control variables are the price $p_i(t)$ and the R&D effort $k_i(t)$, while quality $q_i(t)$ is the state variable. Firms play simultaneously and non-cooperatively. Given the dynamic constraints and the instantaneous profit functions, the resulting Hamiltonian of firm i is not written in a linear-quadratic form, and consequently the feedback solution cannot be obtained analytically through Bellman's equation.[17] Therefore, in characterising the strongly time consistent equilibrium, I will focus on the memoryless closed-loop solution.

10.6.1 The closed-loop solution

Firm i's Hamiltonian function is:

$$\mathcal{H}_i(t) = e^{-\rho t} \cdot \left\{\pi_i(t) + \lambda_{ii}(t)\frac{dq_i(t)}{dt} + \lambda_{ij}(t)\frac{dq_j(t)}{dt}\right\}, \; i, j = H, L; \; i \neq j, \qquad (10.68)$$

where $\lambda_{ij}(t) = \mu_{ij}(t)e^{\rho t}$, and $\mu_{ij}(t)$ is the co-state variable associated to $q_j(t)$. The FOCs w.r.t prices are:

$$\frac{\partial \mathcal{H}_H}{\partial p_H} = \frac{p_L - 2p_H + \Theta(q_H - q_L)}{q_H - q_L} = 0 ; \qquad (10.69)$$

$$\frac{\partial \mathcal{H}_L}{\partial p_L} = \frac{p_H q_L - 2p_L q_H}{q_L(q_H - q_L)} = 0 . \qquad (10.70)$$

[16] To be completely honest, one should also mention a technical reason for this assumption. If δ appeared also in $dq_L(t)/dt$, then the dynamic system would not be analytically solvable.

[17] On the difference between the closed-loop memoryless solution and the feedback solution, see Başar and Olsder (1982, 1995[2], ch. 6; in particular, Proposition 6.1).

These FOCs yield optimal prices:

$$p_H^* = \frac{2\Theta q_H (q_H - q_L)}{4q_H - q_L}\,;\; p_L^* = \frac{\Theta q_L (q_H - q_L)}{4q_H - q_L}\,. \tag{10.71}$$

The remaining conditions are:

$$\frac{\partial \mathcal{H}_i}{\partial k_i} = -2bk_i + a\lambda_{ii} = 0 \Leftrightarrow k_i^* = \frac{a\lambda_{ii}}{2b}\,,\; i = H, L\,; \tag{10.72}$$

$$-\frac{\partial \mathcal{H}_H}{\partial q_H} - \frac{\partial \mathcal{H}_H}{\partial p_L} \cdot \frac{\partial p_L^*}{\partial q_H} - \frac{\partial \mathcal{H}_H}{\partial k_L} \cdot \frac{\partial k_L^*}{\partial q_H} = \frac{\partial \mu_{HH}}{\partial t} \Rightarrow \tag{10.73}$$

$$\frac{\partial \lambda_{HH}}{\partial t} = (\rho + \delta)\,\lambda_{HH} - \frac{p_H (p_H - p_L)}{(q_H - q_L)^2} - \frac{3\Theta p_H q_L^2}{(q_H - q_L)(4q_H - q_L)^2}\,;$$

$$-\frac{\partial \mathcal{H}_H}{\partial q_L} - \frac{\partial \mathcal{H}_H}{\partial p_L} \cdot \frac{\partial p_L^*}{\partial q_L} - \frac{\partial \mathcal{H}_H}{\partial k_L} \cdot \frac{\partial k_L^*}{\partial q_L} = \frac{\partial \mu_{HL}}{\partial t} \tag{10.74}$$

$$-\frac{\partial \mathcal{H}_L}{\partial q_L} - \frac{\partial \mathcal{H}_L}{\partial p_H} \cdot \frac{\partial p_H^*}{\partial q_L} - \frac{\partial \mathcal{H}_L}{\partial k_H} \cdot \frac{\partial k_H^*}{\partial q_L} = \frac{\partial \mu_{LL}}{\partial t} \Rightarrow \tag{10.75}$$

$$\frac{\partial \lambda_{LL}}{\partial t} = \rho \lambda_{LL} - \frac{p_L (p_H q_L^2 - 2p_L q_H q_L + p_L q_H^2)}{[q_L (q_H - q_L)]^2} + \frac{6\Theta p_L q_H^2}{(q_H - q_L)(4q_H - q_L)^2}\,;$$

$$-\frac{\partial \mathcal{H}_L}{\partial q_H} - \frac{\partial \mathcal{H}_L}{\partial p_H} \cdot \frac{\partial p_H^*}{\partial q_H} - \frac{\partial \mathcal{H}_L}{\partial k_H} \cdot \frac{\partial k_H^*}{\partial q_H} = \frac{\partial \mu_{LH}}{\partial t} \tag{10.76}$$

$$\lim_{t \to \infty} \mu_i(t) \cdot q_i(t) = 0\,,\; i = H, L\,. \tag{10.77}$$

First, note that the feedback effects $\frac{\partial u_i^*}{\partial q_j}$ must be calculated on the basis of the optimal levels of $u_i = \{p_i\,,\,k_i\}$ (hence the star) obtaining from the relevant first order conditions. This immediately reveals that

$$\frac{\partial k_i^*}{\partial q_j} = 0 \text{ for all } i, j \tag{10.78}$$

as it can be ascertained from (10.72), and therefore feedback rules are affected only by the effect of state variables on optimal prices, which, solving (10.69-10.70) yields the same prices (10.71) that would be observed in the second stage of the static duopoly game (Choi and Shin, 1992). Second, observe that (10.72) depends upon λ_{ii} only. Consequently, the kinematic equation of k_i is a function of $\partial\lambda_{ii}/\partial t$ but not of $\partial\lambda_{ij}/\partial t$. Therefore, co-state equation (10.74) and (10.76) are redundant and the problem admits the solution $\lambda_{HL} = \lambda_{LH} = 0$. From (10.72) I obtain:

$$\lambda_{ii} = \frac{2bk_i}{a}\,;\; \frac{dk_i}{dt} = \frac{a\lambda_{ii}}{2b} \cdot \frac{\partial \lambda_{ii}}{\partial t}\,. \tag{10.79}$$

Then, using (10.73), (10.75) and (10.79), together with optimal prices $\{p_i^*\}$, I can write:

$$\frac{dk_H}{dt} \propto b(\rho+\delta)(4q_H - q_L)^3 k_H - 2a\Theta^2 q_H \left(4q_H^2 - 3q_H q_L + 2q_L^2\right) ; \quad (10.80)$$

$$\frac{dk_L}{dt} \propto 2b\rho (4q_H - q_L)^3 k_L - a\Theta^2 q_H^2 (4q_H - 7q_L) . \quad (10.81)$$

By imposing $dk_i/dt = 0$, $i = H, L$, I obtain the optimal investment efforts as a function of quality levels:

$$k_H^{CL} = \frac{2a\Theta^2 q_H \left(4q_H^2 - 3q_H q_L + 2q_L^2\right)}{b(\rho+\delta)(4q_H - q_L)^3} ; \quad (10.82)$$

$$k_L^{CL} = \frac{a\Theta^2 q_H^2 (4q_H - 7q_L)}{2b\rho (4q_H - q_L)^3} , \quad (10.83)$$

where superscript CL stands for *closed-loop*.[18] These expressions can be plugged into the co-state equations, to obtain:

$$\frac{dq_H(t)}{dt} = \frac{2a^2\Theta^2 q_H \left(4q_H^2 - 3q_H q_L + 2q_L^2\right)}{b(\rho+\delta)(4q_H - q_L)^3} - \delta q_H ; \quad (10.84)$$

$$\frac{dq_L(t)}{dt} = \frac{a^2\Theta^2 q_H^2 (4q_H - 7q_L)}{b\rho (4q_H - q_L)^3} . \quad (10.85)$$

Now notice that $k_L^{CL} = dq_L(t)/dt = 0$ at $q_L^* = 4q_H/7$, which is the dynamic counterpart of the so-called 4/7 rule derived by Choi and Shin (1992) in correspondence of the two-stage subgame perfect equilibrium of the static game. Using $q_L = 4q_H/7$, (10.84) simplifies as follows:

$$\frac{dq_H(t)}{dt} = \frac{7a^2\Theta^2}{48b(\rho+\delta)} - \delta q_H \quad (10.86)$$

which is equal to zero at:

$$q_H^{CL} = \frac{7a^2\Theta^2}{48b\delta(\rho+\delta)} . \quad (10.87)$$

The optimal low quality level simplifies as follows:

$$q_L^{CL} = \frac{4q_H^{CL}}{7} = \frac{a^2\Theta^2}{12b\delta(\rho+\delta)} , \quad (10.88)$$

[18] The pair $\{k_i^{CL}\}$ is also the unique solution to $dk_i/dt = 0$. Numerical calculations are necessary to show that the difference $k_H^{CL} - k_L^{CL}$ may take either sign, depending on the quality ratio q_H/q_L (out of equilibrium).

showing that the equilibrium level of the low quality also depends on the depreciation rate directly affecting the high quality. The steady state level of investment carried out by firm H is $k_H^{CL} = 7a\Theta^2/[48b\delta(\rho+\delta)]$. Hence, clearly, in equilibrium we have $k_H^{CL} > k_L^{CL} = 0$.

At the closed-loop equilibrium, quantities coincide with those obtained from the static game:

$$x_H^{CL} = \frac{7\Theta}{12} \; ; \; x_L^{CL} = \frac{7\Theta}{24} \, , \tag{10.89}$$

while prices are:

$$p_H^{CL} = \frac{7a^2\Theta^3}{192b\delta(\rho+\delta)} \; ; \; p_L^{CL} = \frac{a^2\Theta^3}{96b\delta(\rho+\delta)} \, . \tag{10.90}$$

Accordingly, partial market coverage holds for all $\Theta \in [1\,,\, 8/7)$. As for equilibrium profits, we have:

$$\pi_H^{CL} = \frac{49a^2\Theta^4\rho}{2304b\delta(\rho+\delta)^2} \; ; \; \pi_L^{CL} = \frac{7a^2\Theta^4}{2304b\delta(\rho+\delta)} \tag{10.91}$$

and:

$$\pi_H^{CL} - \pi_L^{CL} = \frac{7a^2\Theta^4(6\rho-\delta)}{2304b\delta(\rho+\delta)^2} \tag{10.92}$$

with $\pi_H^{CL} > \pi_L^{CL}$ for all $\delta \in [0, 6\rho)$, and conversely outside this parameter range.

The above discussion produces the following:

Theorem 10.1 *Suppose $\Theta \in [1\,,\, 8/7)$. The closed-loop solution of the game entails partial market coverage at $q_L^{CL} = 4q_H^{CL}/7$, with $\pi_L^{CL} > \pi_H^{CL}$ for all $\delta > 6\rho$ (and conversely).*

The properties of the dynamic system can be evaluated at

$$\{q_L^{CL} = 4q_H^{CL}/7\,,\, q_H\,,\, k_H^{CL} = 49a^2\Theta^4/2304\rho^4\,,\, k_L^{CL} = 0\} \tag{10.93}$$

to show that:

Proposition 10.6 *The closed-loop equilibrium is stable. In particular, it is a saddle point.*

Proof. See Appendix 10.1. ■

There remains to assess the duopoly welfare performance at the closed-loop equilibrium. Steady state consumer surplus is:

$$\begin{aligned} CS^{CL} &= CS_H^{CL} + CS_L^{CL}, \text{ i.e.,} \\ CS^{CL} &= \int_{\theta_H}^{\Theta} \left[\theta q_H^{CL} - p_H^{CL}\right] d\theta + \int_{\theta_L}^{\theta_H} \left[\theta q_L^{CL} - p_L^{CL}\right] d\theta = \\ &= \frac{5399a^2\Theta^4}{13824b\delta(\rho+\delta)} + \frac{49a^2\Theta^4}{13824b\delta(\rho+\delta)}. \end{aligned} \quad (10.94)$$

Consequently, social welfare amounts to:

$$SW^{CL} = \pi_H^{CL} + \pi_L^{CL} + CS^{CL} = \frac{7a^2\Theta^4(22\rho+15\delta)}{2304b\delta(\rho+\delta)^2}. \quad (10.95)$$

Now the following question arises. Given that neither the profit-seeking monopolist nor the social planner would supply a second variety, the comparative evaluation of welfare performances across regime can only be carried out taking into account that different product ranges characterise monopoly and duopoly markets. Accordingly, a duopoly may ensure a higher welfare than a single product monopoly, if the trade-off between cost considerations on one side and the output expansion (or average price reduction) on the other is favourable in this respect. Using (10.41) and (10.95), one obtains:

$$SW^{CL} - SW_M^* = \frac{a^2\Theta^4(46\rho+33\delta)}{2304b\delta(\rho+\delta)^2}, \quad (10.96)$$

which is always positive. Then, clearly, we also have that $SW_{SP}^* > SW^{CL}$ always. Of course these inequalities come as no surprise, but their ultimate implications are non-trivial. To see this, consider an extension of the present model where (at least unilateral) international trade is accounted for. Suppose there are two countries, each initially supplied by a single firm in autarky. As we know from the above analysis, such a firm will surely be a single-product one. The government of each country is aware of this fact, i.e., that the autarkic monopolist will not supply a second variety. Therefore, the government may eliminate any barriers to trade for the sake of having another variety being supplied by the foreign firm to the domestic market. This will be welfare-improving, even if some domestic surplus is going to accrue to the foreign country in the form of its firm's profits. This can be quickly proved, under the assumption that the support $[\Theta - 1, \Theta]$ of the distribution of the marginal willingness to pay is the same in both countries, and there is one-way trade from the foreign country to the domestic market. In such a case, the above calculations for the closed-loop duopoly equilibrium also apply to the international trade setting.

Assume first that the home firm is the high-quality supplier, so that imports are x_L^{CL}. The social welfare enjoyed by the home country at equilibrium is:

$$SW_H^{CL} = \pi_H^{CL} + CS_H^{CL} + CS_L^{CL} = \frac{49a^2\Theta^4 (3\rho + 2\delta)}{2304b\delta (\rho + \delta)^2} > SW_M^* . \qquad (10.97)$$

If instead the domestic producer is the low-quality firm when trade opens, then we have:

$$SW_L^{CL} = \pi_L^{CL} + CS_H^{CL} + CS_L^{CL} = \frac{35a^2\Theta^4}{2304b\delta (\rho + \delta)} , \qquad (10.98)$$

with:

$$SW_L^{CL} - SW_M^* = \frac{a^2\Theta^4 (11\delta - \rho)}{768b\delta (\rho + \delta)^2} > 0 \text{ for all } \rho \in [0, 11\delta) , \qquad (10.99)$$

which is very likely to be true, as common sense suggests.[19]

Here I do not dwell upon proper (two-way) intraindustry trade, as it is a matter that goes beyond the scope of the present chapter. However, this is a relevant and promising line for future research.

10.7 Concluding remarks

I have analysed two different versions of a differential game where firms, through capital accumulation over time, supply vertically differentiated goods. The explicit treatment of R&D activity as a capital accumulation process proves that several results which are seemingly well established in the static approach are not robust. In particular, I have shown four main results: (i) the sustainability of the duopoly regime is conditional upon the level of firms' R&D investments; (ii) the high quality firm always invests more than the low quality firm; (iii) there exist quality ranges where the low quality firm's profits are larger than the high quality firm's; (iv) at the closed-loop equilibrium, the optimal quality ratio is always 4/7, as in the static model with costless quality improvements (Choi and Shin, 1992).

In consideration of the large number of economically relevant issues associated with the supply of product quality in competitive environments, the

[19] Usually, ρ is taken to be lower than δ, but this consideration is made concerning the depreciation of *capital*, or productive *capacity*. Remember that, here, δ is the depreciation/obsolescence rate affecting the high quality, as perceived by consumers' eyes.

foregoing analysis represents a preliminary step. Many extensions, e.g., dynamic investment in demand-increasing advertising or the accumulation of capacity for production, are left for future research.

Appendix 10.1: Proof of Proposition 10.6

Given the quasi-static solution for prices, the relevant differential equations are the state equations:

$$\begin{aligned} \frac{dq_H(t)}{dt} &= ak_H(t) - \delta q_H(t) ; \\ \frac{dq_L(t)}{dt} &= ak_L(t) . \end{aligned}$$

and the equations of motions of the R&D investment efforts:

$$\frac{dk_H}{dt} = \frac{b(\rho+\delta)(4q_H - q_L)^3 k_H - 2a\Theta^2 q_H (4q_H^2 - 3q_H q_L + 2q_L^2)}{b(4q_H - q_L)^3} ;$$

$$\frac{dk_L}{dt} = \frac{2b\rho(4q_H - q_L)^3 k_L - a\Theta^2 q_H^2 (4q_H - 7q_L)}{2b\rho(4q_H - q_L)^3} .$$

Given that $\lambda_{ij} = 0$ for $i \neq j$, the properties of the system can be established on the basis of the sign of the trace and determinant of the following Jacobian matrix:

$$\Xi_i = \left[\begin{array}{cc} \dfrac{\partial \dot{q}_i}{\partial q_i} & \dfrac{\partial \dot{q}_i}{\partial k_i} \\ \dfrac{\partial \dot{k}_i}{\partial q_i} & \dfrac{\partial \dot{k}_i}{\partial k_i} \end{array} \right] \Bigg|_{q_L = 4q_H/7, k_i^{CL}}$$

for each firm, at

$$\left\{ q_L^{CL} = 4q_H^{CL}/7 \, , \, q_H \, , \, k_H^{CL} = 49a^2\Theta^4/2304\rho^4 \, , \, k_L^{CL} = 0 \right\} .$$

Consider first firm H :

$$\begin{aligned} \frac{\partial \dot{q}_H}{\partial q_H} &= -\delta \, ; \, \frac{\partial \dot{q}_H}{\partial k_H} = a \, ; \\ \frac{\partial \dot{k}_H}{\partial q_H} &= \frac{4a\Theta^2 q_L^2 (5q_H + q_L)}{b(4q_H - q_L)^4} \, ; \, \frac{\partial \dot{k}_H}{\partial k_H} = \rho + \delta \, . \end{aligned}$$

The trace and determinant of Ξ_H are:

$$Tr\left(\Xi_H\right) = \rho > 0 \; ; \; \Delta\left(\Xi_H\right) = -\frac{4a^2\Theta^2 q_L^2 \left(5q_H + q_L\right)}{b\left(4q_H - q_L\right)^4} - \delta\left(\rho + \delta\right) < 0 \, .$$

Examine now firm L :

$$\begin{aligned} \frac{\partial \dot{q}_L}{\partial q_L} &= 0 \, ; \, \frac{\partial \dot{q}_L}{\partial k_L} = a \, ; \\ \frac{\partial \dot{k}_L}{\partial q_L} &= \frac{a\Theta^2 q_H^2 \left(8q_H + 7q_L\right)}{b\left(4q_H - q_L\right)^4} \, ; \, \frac{\partial \dot{k}_H}{\partial k_H} = \rho \, . \end{aligned}$$

The trace and determinant of Ξ_L are:

$$Tr\left(\Xi_L\right) = 2\rho > 0 \; ; \; \Delta\left(\Xi_L\right) = -\frac{a^2\Theta^2 q_H^2 \left(8q_H + 7q_L\right)}{b\left(4q_H - q_L\right)^4} < 0 \, .$$

Therefore, the system evolves along a stable (saddle) path, and the closed-loop equilibrium is a saddle point. ■

References

1] Aoki, R. and T. Prusa (1997), "Sequential versus Simultaneous Choice with Endogenous Quality", *International Journal of Industrial Organization*, **15**, 103-21.

2] Başar, T. and G.J. Olsder (1982, 1995²), *Dynamic Noncooperative Game Theory*, San Diego, Academic Press.

3] Beath, J., Y. Katsoulacos and D. Ulph (1987), "Sequential Product Innovation and Industry Evolution", *Economic Journal*, **97**, 32-43.

4] Cellini, R. and L. Lambertini (1998), "A Dynamic Model of Differentiated Oligopoly with Capital Accumulation", *Journal of Economic Theory*, **83**, 145-55.

5] Cellini, R. and L. Lambertini (2001), "Differential Oligopoly Games where the Closed-Loop Memoryless and Open-Loop Equilibria Coincide", working paper no. 402, Dipartimento di Scienze Economiche, Università degli Studi di Bologna.

6] Cellini, R. and L. Lambertini (2003a), "Differential Oligopoly Games", in P. Bianchi and L. Lambertini (eds), *Technology, Information and Market Dynamics: Topics in Advanced Industrial Organization*, Cheltenham, Edward Elgar, 173-207.

7] Cellini, R. and L. Lambertini (2003b), "Advertising in a Differential Oligopoly Game", *Journal of Optimization Theory and Applications*, **116**, 61-81.
8] Champsaur, P., and J.-C. Rochet (1989), "Multiproduct Duopolists," *Econometrica*, **57**, 533-57.
9] Chiang, A.C. (1992), *Elements of Dynamic Optimization*, New York, McGraw-Hill.
10] Choi, J.C. and H.S. Shin (1992), "A Comment on a Model of Vertical Product Differentiation", *Journal of Industrial Economics*, **40**, 229-31.
11] Clemhout, S. and H.Y. Wan, Jr. (1994), "Differential Games. Economic Applications", in R.J. Aumann and S. Hart (eds.), *Handbook of Game Theory*, Amsterdam, North-Holland, vol. 2, ch. 23, 801-25.
12] Colombo, L. and L. Lambertini (2003), "Dynamic Advertising under Vertical Product Differentiation", *Journal of Optimization Theory and Applications*, **119**, 261-80.
13] Conrad, K. (1985), "Quality, Advertising and the Formation of Goodwill under Dynamic Conditions", in G. Feichtinger (ed.), *Optimal Control Theory and Economic Analysis*, vol. 2, Amsterdam, North-Holland, 215-34.
14] Cremer, H. and J.-F. Thisse (1994), "Commodity Taxation in a Differentiated Oligopoly", *International Economic Review*, **35**, 613-33.
15] Dockner, E.J, S. Jørgensen, N. Van Long and G. Sorger (2000), *Differential Games in Economics and Management Science*, Cambridge, Cambridge University Press.
16] Dutta, P.K., S. Lach and A. Rustichini (1995), "Better Late than Early: Vertical Differentiation in the Adoption of a New Technology", *Journal of Economics and Management Strategy*, **4**, 563-89.
17] Eisner, R. and R.H. Strotz (1963), "Determinants of Business Investment", in *Impacts of Monetary Policy*, Research Studies Prepared for the Commission on Money and Credit, Englewood Cliffs, NJ, Prentice-Hall.
18] Erickson, G.M. (1991), *Dynamic Models of Advertising Competition*, Dordrecht, Kluwer.
19] Evans, G.C. (1924), "The Dynamics of Monopoly", *American Mathematical Monthly*, **31**, 75-83.
20] Feichtinger, G. (1983), "The Nash Solution of an Advertising Differential Game: Generalization of a Model by Leitmann and Schmitendorf", *IEEE Transactions on Automatic Control*, **28**, 1044-1048.
21] Feichtinger, G. and S. Jørgensen (1983), "Differential Game Models in Management Science", *European Journal of Operational Research*, **14**, 137-155.
22] Feichtinger, G., R.F. Hartl and P.S. Sethi (1994), "Dynamic Optimal Control Models in Advertising: Recent Developments", *Management Science*, **40**, 195-226.

23] Fershtman, C. and E. Muller (1984), "Capital Accumulation Games of Infinite Duration", *Journal of Economic Theory*, **33**, 322-39.
24] Fudenberg, D. and J. Tirole (1983), "Capital as a Commitment: Strategic Investment to Deter Mobility", *Journal of Economic Theory*, **31**, 227-50.
25] Gabszewicz, J.J. and J.-F. Thisse (1979), "Price Competition, Quality and Income Disparities", *Journal of Economic Theory*, **20**, 340-59.
26] Gabszewicz, J.J. and J.-F. Thisse (1980), "Entry (and Exit) in a Differentiated Industry", *Journal of Economic Theory*, **22**, 327-38.
27] Jørgensen, S. (1982), "A Survey of Some Differential Games in Advertising", *Journal of Economic Dynamics and Control*, **4**, 341-69.
28] Kotowitz, Y. and F. Mathewson (1979), "Advertising, Consumer Information, and Product Quality", *Bell Journal of Economics*, **10**, 566-88.
29] Lambertini, L. (1996), "Choosing Roles in a Duopoly for Endogenously Differentiated Products", *Australian Economic Papers*, **35**, 205-24.
30] Lambertini, L. (1999), "Endogenous Timing and the Choice of Quality in a Vertically Differentiated Duopoly", *Research in Economics* (*Ricerche Economiche*), **53**, 101-9.
31] Lambertini, L. (2000), "Technology and Cartel Stability under Vertical Differentiation", *German Economic Review*, **1**, 421-42.
32] Lambertini, L., S. Poddar and D. Sasaki (2002), "Research Joint Ventures, Product Differentiation and Price Collusion", *International Journal of Industrial Organization*, **20**, 829-54.
33] Lehmann-Grube, U. (1997), "Strategic Choice of Quality when Quality is Costly: The Persistence of the High-Quality Advantage", *RAND Journal of Economics*, **28**, 372-84.
34] Leitmann, G. and W.E. Schmitendorf (1978), "Profit Maximization through Advertising: A Nonzero Sum Differential Game Approach", *IEEE Transactions on Automatic Control*, **23**, 646-50.
35] Moorthy, K.S. (1988), "Product and Price Competition in a Duopoly Model", *Marketing Science*, **7**, 141-68.
36] Motta, M. (1992), "Cooperative R&D and Vertical Product Differentiation", *International Journal of Industrial Organization*, **10**, 643-61.
37] Motta, M. (1993), "Endogenous Quality Choice: Price vs Quantity Competition", *Journal of Industrial Economics*, **41**, 113-32.
38] Mussa, M., and S. Rosen (1978), "Monopoly and Product Quality," *Journal of Economic Theory*, **18**, 301-17.
39] Nerlove, M. and K.J. Arrow (1962), "Optimal Advertising Policy under Dynamic Conditions", *Economica*, **29**, 129-42.
40] Reynolds, S.S. (1987), "Capacity Investment, Preemption and Commitment in an Infinite Horizon Model", *International Economic Review*, **28**, 69-88.

41] Reynolds, S.S. (1991), "Dynamic Oligopoly with Capacity Adjustment Costs", *Journal of Economic Dynamics and Control*, **15**, 491-514.
42] Ringbeck, J. (1985), "Mixed Quality and Advertising Strategies under Asymmetric Information", in G. Feichtinger (ed.), *Optimal Control Theory and Economic Analysis*, vol. 2, Amsterdam, North-Holland, 197-214.
43] Rosenkranz, S. (1995), "Innovation and Cooperation under Vertical Product Differentiation", *International Journal of Industrial Organization*, **13**, 1-22.
44] Rosenkranz, S. (1997), "Quality Improvements and the Incentive to Leapfrog", *International Journal of Industrial Organization*, **15**, 243-61.
45] Sethi, S.P. (1977), "Dynamic Optimal Control Models in Advertising: A Survey", *SIAM Review*, **19**, 685-725.
46] Shaked, A. and J. Sutton (1982), "Relaxing Price Competition through Product Differentiation", *Review of Economic Studies*, **69**, 3-13.
47] Shaked, A. and J. Sutton (1983), "Natural Oligopolies", *Econometrica*, **51**, 1469-83.
48] Singh, N. and X. Vives (1984), "Price and Quantity Competition in a Differentiated Duopoly", *RAND Journal of Economics*, **15**, 546-54.
49] Spence, A.M. (1975), "Monopoly, Quality and Regulation," *Bell Journal of Economics*, **6**, 417-29.
50] Tintner, G. (1937), "Monopoly over Time", *Econometrica*, **5**, 160-70.
51] Tirole, J. (1988), *The Theory of Industrial Organization*, Cambridge, MA, MIT Press.
52] Vidale, M.L. and H.B. Wolfe (1957), "An Operations Research Study of Sales Response to Advertising", *Operations Research*, **5**, 370-81.

Chapter 11

A differential game of advertising under vertical differentiation

Luca Colombo and Luca Lambertini

11.1 Introduction

In this final chapter, we aim at investigating a dynamic advertising model under vertical differentiation.[1] The scanty literature currently available in this field usually considers advertising as an instrument to convey information about the existence and the characteristics of the advertised good or as a way to increase the stock of goodwill or reputation (Kotowitz and Mathewson, 1979; Conrad, 1985; Ringbeck, 1985). Two exceptions dealing with a full information setting are in Ouardighi and Pasin (2002), extending the well known Lanchester model to account for the interplay between market shares and quality, and in Colombo and Lambertini (2003), assuming a direct relationship between advertising efforts and the rate of change in sales, as in Vidale and Wolfe (1957), under endogenous vertical differentiation.[2] Others have investigated, adopting either static or dynamic approaches, the strategic use of product qualities as firms' instruments to build up market shares (Moorthy, 1988; Motta, 1993; Dutta *et al.*, 1995).

In line of principle, the informative role played by advertising is plausible

[1] For exhaustive surveys on dynamic advertising, see Sethi (1977); Jørgensen (1982); Erickson (1991); Feichtinger *et al.* (1994).

[2] For the formulation of the Lanchester model, see Dockner *et al.* (2000, ch. 11), Case (1979), Sorger (1989) and Erickson (1991).

only in a world where the dissemination of information is scarce and where some product characteristics are not perfectly observable by consumers before purchase. Many goods that are actually advertised in modern economies, however, seem to be at odds with this world. Examples abound. Indeed, when the informative gap between producers and consumers is either negligible or, whatever it is, it cannot be reduced by advertising, the informative and signaling role of advertising becomes so marginal that it can not be justified from a theoretical point of view. For instance, the exact formula of Coca Cola is still unknown and, presumably, so it will remain in the foreseeable future. Accordingly, the focus of Coca Cola's advertising campaigns is not information. In this context, the unique theoretical explanation of the fact that huge amounts of money are spent by firms in advertising campaigns, is that either advertising acts so as to increase the marginal willingness to pay of consumers for the advertised good, or the consequence of a prisoners' dilemma generated by the presence of rival firms (like Pepsi) adopting similar strategies. The subject matter of this chapter is the first of the two possibilities.[3] Having said that, a natural question should come into mind: why perfectly informed (and perfectly rational) consumers should be ready to pay more for a good the characteristics of which, including the price vector, are common knowledge? The answer is that even if all the physical characteristics of the good remain unchanged by the fact of being advertised, the psychological responses of consumers are not.

As perfectly recognized by Galbraith (1967, ch. XVIII), advertising in modern economies is much more concerned with being persuasive rather than informative, where being persuasive means being able to add extra values to the advertised good by conferring to its owner a sense of personal achievement, social recognition, beauty, by diverting his mind from thought, or by being in whatever other manner psychologically rewarding.

In Galbraith's vein, we assume that advertising makes perfectly informed consumers more willing to pay for the good being advertised. We investigate a differential game in order to study the dynamic incentives for oligopolistic firms to invest in such persuasive advertising campaigns coupled with product quality improvements, which are the result of capital accumulation over time. For the sake of expositional homogeneity, we adopt the same structure of consumers' preferences as the one which has been used throughout the book. We characterize both open-loop and closed-loop memoryless Nash equilibria, and proceed to a steady state (saddle path) analysis. Our main results can be summarized as follows: in line with the existing static literature on product

[3]For the analysis of the first perspective, where advertising aims at attracting additional consumers, see Colombo and Lambertini (2003).

quality provision in oligopoly (Gabszewicz and Thisse, 1979, 1980; Shaked and Sutton, 1982, 1983; Lehmann-Grube, 1997, *inter alia*), we show that the high quality firm serves more customers and invests more than the low quality firm. However, the resulting ranking of firms' profits may significantly differ from the one we are accustomed to from the aforementioned literature. More precisely, there exists an admissible subset of parameters wherein the low quality firm performs better than the high quality firm in terms of equilibrium profits. Moreover, we prove that it is possible for the low quality firm to become a monopolist, provided that future profits sufficiently matter.

The remainder of this chapter is structured as follows. The model is described in section 11.2. Section 11.3 deals with the solution of the game under both open-loop and closed-loop memoryless solution concept. Concluding remarks are in section 11.4.

11.2 The Model

Time is continuous and indicated by t. At each $t \in [0, \infty)$, a market for vertically differentiated goods exists. Let this market be supplied by two single-product firms offering goods of quality $q_i(t)$ at a price $p_i(t)$, with $i = H, L$, with $\infty > q_H(t) \geq q_L(t) \geq 0$ at any t. Production exhibits constant returns to scale and, without any loss of generality, we normalize the unit production cost to zero.

On the demand side, each consumer is characterized by a marginal willingness to pay for quality θ, uniformly distributed over the support $[\Theta(t) - 1, \Theta(t)]$, with $\Theta(t) > 1$ at any t. We assume $f(\theta) = 1$, implying that consumers' population is normalized to 1.[4] For the sake of simplicity, we abstract from the presence of switching costs, so customers that, as time goes by, switch from one variety to the other, bear any disutility.[5]

The instantaneous net surplus a consumer of type θ draws from the variety characterized by $q_i(t)$ is defined as follows:

$$U_\theta(t) = \begin{cases} \theta q_i(t) - p_i(t) \geq 0 & \text{if he buys variety } i = H, L \\ 0 & \text{if he doesn't buy} \end{cases} \tag{11.1}$$

In order to derive the expressions of market demands, we compute the threshold value of θ characterising the consumer who is indifferent between

[4] At each point in time, each consumer buys at most one unit of the preferred variety. This rules out the use of second-degree price discrimination.

[5] For an exhaustive survey on consumers' switching costs, see Klemperer (1995).

buying the high quality good and buying the low quality good:

$$\widehat{\theta}(t) = \frac{p_H(t) - p_L(t)}{q_H(t) - q_L(t)}, \tag{11.2}$$

and the one which characterizes the consumer who is indifferent between buying the low quality good and not buying at all:

$$\widetilde{\theta}(t) = \frac{p_L(t)}{q_L(t)}. \tag{11.3}$$

Accordingly, the direct demand system obtains:

$$x_H(t) = \Theta(t) - \widehat{\theta}(t) ; \tag{11.4}$$

$$x_L(t) = \widehat{\theta}(t) - \widetilde{\theta}(t) , \tag{11.5}$$

which can be inverted as long as partial market coverage prevails, i.e., for all $\widetilde{\theta}(t) > 0$:

$$p_H(t) = q_H(t)[\Theta(t) - x_H(t)] - q_L(t)x_L(t) ; \tag{11.6}$$

$$p_L(t) = q_L(t)[\Theta(t) - x_H(t) - x_L(t)] . \tag{11.7}$$

Firm i's instantaneous profits write:

$$\pi_H(t) = p_H(t)x_H(t) - [a_H(t)]^2 - [b_H(t)]^2 ; \tag{11.8}$$

$$\pi_L(t) = p_L(t)x_L(t) - [a_L(t)]^2 - [b_L(t)]^2 , \tag{11.9}$$

where $[a_i(t)]^2$ and $[b_i(t)]^2$ are the instantaneous quadratic costs associated with persuasive advertising campaigns and product quality improvements, respectively; $a_i(t)$ denotes the instantaneous investments in persuasive advertising campaigns made by firm i at time t, and $b_i(t)$ denotes the instantaneous R&D investments in product quality improvements made by firm i at time t.

We assume that the quality of firm i's product evolves over time according to the following dynamics:

$$\frac{dq_i(t)}{dt} \equiv \dot{q_i} = b_i(t) - \delta q_i(t), \quad i = \{H, L\} , \tag{11.10}$$

while the upper bound of the support in which the marginal willingness to pay for quality lies evolves over time according to the following dynamics:

$$\frac{d\Theta(t)}{dt} \equiv \dot{\Theta} = a_H(t) + a_L(t) - \delta\Theta(t) , \tag{11.11}$$

where $\delta > 0$ denotes the common depreciation rate.[6] It is worth noting the differences in structure between (11.10) and (11.11): while the quality of firm i's product evolves over time independently of firm j's product quality and investment, the dynamics of Θ features non rival properties which are typical of public goods. In this respect, the effects induced by individual advertising campaigns consist in making not only own consumers, but all the consumers in the market more quality-oriented, including those whose willingness to pay for quality is still too low to buy.[7] It is also interesting to compare (11.10) with its analogous counterpart of the last chapter. There, quality depreciation was assumed to be nil for firm L, while here we assume that quality depreciates at the rate δ for both firms.

The objective of firm i consists in maximising the present value of its profits stream over an infinite time horizon w.r.t. control variables $a_i(t)$, $b_i(t)$ and $x_i(t)$, under the constraint given by states dynamics:

$$\begin{aligned} &\max_{a_i(t),b_i(t),x_i(t)} \Pi_i(t) = \int_0^{\infty} \pi_i(t)e^{-\rho t}dt \\ &\text{s.t. } \frac{dq_i(t)}{dt} \equiv \dot{q}_i = b_i(t) - \delta q_i(t), \quad i = \{H, L\} \\ &\text{and } \frac{d\Theta(t)}{dt} \equiv \dot{\Theta} = a_H(t) + a_L(t) - \delta\Theta(t) . \end{aligned} \tag{11.12}$$

The discount rate $\rho > 0$ is assumed to be constant and common to both firms.

11.3 The game

Firm i's current value Hamiltonian writes:

$$\mathcal{H}_i(t) = e^{-\rho t} \cdot \left\{ \pi_i(t) + \lambda_{ii}(t)\dot{q}_i + \lambda_{ij}(t)\dot{q}_j + \nu_i(t)\dot{\Theta}(t) \right\}, \tag{11.13}$$

where $\lambda_{ii}(t) = \mu_{ii}(t)e^{\rho t}$, $\lambda_{ij}(t) = \mu_{ij}(t)e^{\rho t}$ and $\nu_i(t) = \kappa_i(t)e^{\rho t}$; $\mu_{ii}(t)$ and $\mu_{ij}(t)$ are the co-state variable associated to $q_i(t)$ and $\kappa_i(t)$ is the co-state variable associated to $\Theta(t)$. Firms play simultaneously in each point in time. First order conditions (FOCs) on controls are (henceforth, the indication of

[6] The assumption that parameter δ is the same for both quality and the marginal willingness to pay is not at all crucial, and has been made only to end up with more tractable solutions than otherwise.

[7] A differential game where firms' advertising campaigns have a public good nature is in Cellini and Lambertini (2003).

time and exponential discounting are omitted for brevity):[8]

$$\frac{\partial \mathcal{H}_H}{\partial x_H} = \Theta q_H - 2q_H x_H - q_L x_L = 0 \; ; \tag{11.14}$$

$$\frac{\partial \mathcal{H}_L}{\partial x_L} = q_L \left(\Theta - x_H - 2x_L \right) = 0 \; ; \tag{11.15}$$

$$\frac{\partial \mathcal{H}_i}{\partial a_i} = -2a_i + \lambda_{ii} = 0 \; , \; i = H, L \; ; \tag{11.16}$$

$$\frac{\partial \mathcal{H}_i}{\partial b_i} = -2b_i + \lambda_{ii} = 0 \; , \; i = H, L \; . \tag{11.17}$$

The above FOCs, in particular equations (11.14) and (11.15), imply that, by applying the open-loop solution concept to the present game, we end up with equilibria which are not subgame perfect.[9] As argued in the previous chapter, this can be justified by considering that, in some circumstances, it may be too costly for firms to modify their investment plans on the way. Notice also that FOCs do not contain λ_{ij} because of the assumptions concerning the state equations, characterised by separated dynamics. Therefore, we set $\lambda_{ij} = 0$ for all $t \in [0, \infty)$ and $j \neq i$, and specify only two co-state equations per firm, disregarding the one pertaining to the rival's quality. We first solve the game in the open-loop form. Then, in order to characterize equilibria that are strongly time consistent, we will solve the game according to the closed-loop memoryless solution concept.

11.3.1 Open-Loop Equilibrium

Under the open-loop solution concept, by definition, feedback effects are not taken into account. The relevant co-state equations write as follows:

$$\begin{aligned} -\frac{\partial \mathcal{H}_i}{\partial q_i} &= \frac{\partial \lambda_{ii}}{\partial t} - \rho \lambda_{ii} \Leftrightarrow \\ \frac{\partial \lambda_{ii}}{\partial t} &= \left(\rho + \delta \right) \lambda_{ii} - \left(\Theta - x_H \right) x_H \end{aligned} \tag{11.18}$$

$$\begin{aligned} -\frac{\partial \mathcal{H}_i}{\partial \Theta} &= \frac{\partial \nu_i}{\partial t} - \rho \nu_i \Leftrightarrow \\ \frac{\partial \nu_i}{\partial t} &= \left(\rho + \delta \right) \mu_i - q_H x_H \end{aligned} \tag{11.19}$$

[8] Second order conditions are met throughout the paper. They are omitted for brevity.

[9] See, e.g., Mehlman and Willing (1983), Reinganum (1982), Dockner, Feichtinger and Jørgensen (1985) and Fershtman (1987). For an exhaustive discussion on the coincidence between open-loop and closed-loop memoryless solutions, see Dockner *et al.* (2000, ch.7).

along with the transversality conditions:

$$\lim_{t\to\infty} \mu_i(t) \cdot q_i(t) = 0 \text{ and } \lim_{t\to\infty} \kappa_i(t) \cdot \Theta(t) = 0,\ i = H, L \tag{11.20}$$

and the initial conditions

$$q_i(0) = q_{i0} > 0, \text{ with } q_{H0} > q_{L0} \text{ and } \Theta(0) = \Theta_0 \geq 1. \tag{11.21}$$

Solving (11.14-11.15) yields the equilibrium output levels for a generic quality pair:

$$x_H^* = \Theta\left(1 - \frac{2q_H}{4q_H - q_L}\right) ; \ x_L^* = \frac{\Theta q_H}{4q_H - q_L}, \tag{11.22}$$

with $x_H^* > x_L^* > 0$ for all $q_H > q_L > 0$. If $q_H = q_L$, $p_H^* = p_L^* = 0$ and the allocation of market demand across firms is not determined. Notice that, since output levels do not appear in the state dynamics, the optimal solution obtained through the Hamiltonian coincides with the static ones (see section 2 of the previous chapter).

From (11.16) and (11.17) we obtain:

$$\lambda_{ii} = 2a_i \Rightarrow \dot{\lambda}_{ii} = 2\dot{a}_i; \tag{11.23}$$

$$\mu_{ii} = 2b_i \Rightarrow \dot{\mu}_{ii} = 2\dot{b}_i. \tag{11.24}$$

Now, by plugging (11.23) into (11.18) and (11.24) into (11.19), and using (11.22), we derive the dynamics of firm H's investments:

$$\dot{a}_H = a_H(\delta + \rho) + \frac{\Theta q_H (q_L - 2q_H)}{8q_H - 2q_L}; \tag{11.25}$$

$$\dot{b}_H = b_H(\delta + \rho) + \frac{\Theta^2 q_H (q_L - 2q_H)}{(q_L - 4q_H)^2}, \tag{11.26}$$

and those referred to firm L:

$$\dot{a}_L = a_L(\delta + \rho) - \frac{\Theta q_H q_L}{8q_H - 2q_L}; \tag{11.27}$$

$$\dot{b}_L = b_L(\delta + \rho) - \frac{\Theta^2 q_H^2}{2(q_L - 4q_H)^2}. \tag{11.28}$$

The dynamic system formed by (11.25), (11.26), (11.27) and (11.28), together with the state equations (11.10) and (11.11), yields the following admissible steady state point:[10]

$$a_H^{OL} = 4.6629\delta^2(\delta + \rho) \ ; \ a_L^{OL} = 0.80002\delta^2(\delta + \rho) \ ; \tag{11.29}$$

[10] We find four equilibria, only the one reported in the text being admissible. Among the other three critical points, two are non real and one does not respect the condition $q_H > q_L$.

$$b_H^{OL} = 3.7071\delta^2 (\delta + \rho) \; ; \; b_L^{OL} = 1.0858\delta^2 (\delta + \rho) \; ; \tag{11.30}$$

$$\Theta^{OL} = 5.46298\delta (\delta + \rho) \; ; \; q_H^{OL} = 3.7071\delta (\delta + \rho) \; ; \; q_L^{OL} = 1.0858\delta (\delta + \rho) \; . \tag{11.31}$$

Proposition 11.1 *The steady state identified by*

$$\{a_H^{OL}, a_H^{OL}, b_H^{OL}, b_L^{OL}, \Theta^{OL}, q_H^{OL}, q_L^{OL}\}$$

is a saddle point equilibrium.

Proof. See Appendix 11.1. ■

By a direct comparison between equilibrium qualities and investments, we can write:

Lemma 11.1 *Under the open-loop solution concept, firm H invests more than firm L both in advertising and in quality improvement.*

Concerning equilibrium output levels, we have:

$$x_H^{OL} = 2.5156\delta (\delta + \rho) \; ; \; x_L^{OL} = 1.4736\delta (\delta + \rho) \; , \tag{11.32}$$

which are admissible iff $x_H^{OL} + x_L^{OL} < 1$. This condition is satisfied iff:

$$\rho < \overline{\rho} = -1.0027 \times 10^{-4} \frac{-2500 + 9973\delta^2}{\delta} \; . \tag{11.33}$$

Equilibrium prices are:

$$p_H^{OL} = 9.3259\delta^2 (\delta + \rho)^2 \; ; \; p_L^{OL} = 1.6001\delta^2 (\delta + \rho)^2 \; . \tag{11.34}$$

Lemma 11.2 *Under the open-loop solution concept, firm H attains a larger market share and charges a higher market price than firm L.*

We are now in a position to assess the relative performance of firms in terms of equilibrium profits:

$$\pi_H^{OL} = 23.46\delta^3 (\delta + \rho)^3 - 35.485\delta^4 (\delta + \rho)^2 \; ; \tag{11.35}$$

$$\pi_L^{OL} = 2.357\,9\delta^3 (\delta + \rho)^3 - 1.819\delta^4 (\delta + \rho)^2 \; . \tag{11.36}$$

The sustainability of either the monopoly or the duopoly regime depends upon the non-negativity of profits, which, in turns, depends upon intertemporal parameters. In this respect, our main result is as follows:

Proposition 11.2 *Provided that $\rho \in [0, \overline{\rho})$, ensuring that partial market coverage prevails, under the open-loop solution concept the following holds:*

(i) for all $\rho \in [0, 0.51257\delta)$, we have $\pi_L^{OL} > 0 > \pi_H^{OL}$;

(ii) for all $\rho \in (0.51257\delta, 0.59539\delta]$, we have $\pi_L^{OL} > \pi_H^{OL} > 0$;

(iii) for all $\rho > 0.59539\delta$, we have $\pi_H^{OL} > \pi_L^{OL} > 0$.

The following figure illustrates the above proposition:

Figure 11.1 : Parameter space

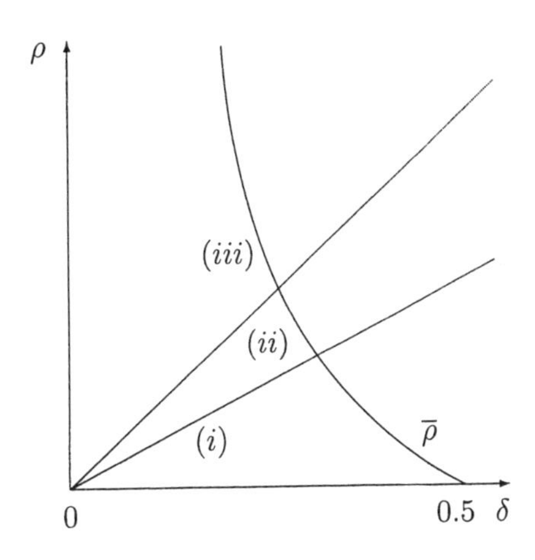

In region (i), only the low quality firm survives. In region (ii), the market is served by both firms; yet, contrary to the conventional wisdom coming from the existing static literature, the low quality firm outperforms the high quality firm in terms of equilibrium profits. Finally, in region (iii), we find the traditional (or *quasi*-static) result on the distribution of profits in a setting where quality improvements require capital accumulation to take place and consumers' preferences are affected by firms' advertising campaigns carried out over time. This is in accordance with the fact that, as the discount rate becomes higher, firms become increasingly myopic, and the dynamic game is perceived as closer to the static one.

11.3.2 Closed-Loop Memoryless Equilibrium

Now, we write firm i's co-state equations according to the closed-loop memoryless solution concept:

$$-\frac{\partial \mathcal{H}_i}{\partial q_i} - \frac{\partial \mathcal{H}_i}{\partial x_j}\frac{\partial x_j^*}{\partial q_i} - \frac{\partial \mathcal{H}_i}{\partial a_j}\frac{\partial a_j^*}{\partial q_i} - \frac{\partial \mathcal{H}_i}{\partial b_j}\frac{\partial b_j^*}{\partial q_i} = \frac{\partial \lambda_{ii}}{\partial t} - \rho\lambda_{ii} \; ; \tag{11.37}$$

$$-\frac{\partial \mathcal{H}_i}{\partial \Theta} - \frac{\partial \mathcal{H}_i}{\partial x_j}\frac{\partial x_j^*}{\partial \Theta} - \frac{\partial \mathcal{H}_i}{\partial a_j}\frac{\partial a_j^*}{\partial \Theta} - \frac{\partial \mathcal{H}_i}{\partial b_j}\frac{\partial b_j^*}{\partial \Theta} = \frac{\partial \nu_i}{\partial t} - \rho\nu_i \; , \tag{11.38}$$

where starred variables indicate that partial derivatives are obtained using FOCs, along with the same transversality and initial conditions as in open-loop. First, notice that:

$$\frac{\partial a_j^*}{\partial q_i} = \frac{\partial b_j^*}{\partial q_i} = 0 \; ; \; \frac{\partial a_j^*}{\partial \Theta} = \frac{\partial b_j^*}{\partial \Theta} = 0 \; , \tag{11.39}$$

meaning that the only feedbacks that have to be taken into account are those regarding optimal output levels:

$$\frac{\partial \mathcal{H}_i}{\partial x_j} = q_L x_i \; ; \; \frac{\partial x_H^*}{\partial q_L} = -\frac{x_L}{2q_H} \; ; \; \frac{\partial x_L^*}{\partial q_H} = 0 \; ; \; \frac{\partial x_j^*}{\partial \Theta} = \frac{1}{2} \, . \tag{11.40}$$

Now, by plugging (11.23) into (11.37) and (11.24) into (11.38), and using equilibrium sales (11.22), we derive the dynamics of firm H's investments:

$$\dot{a}_H = a_H (\delta + \rho) + \Theta \left(\frac{q_L}{4} - \frac{q_H^2}{4q_H - q_L} \right) \; ; \tag{11.41}$$

$$\dot{b}_H = b_H (\delta + \rho) + \frac{\Theta^2 q_H (q_L - 2q_H)}{(q_L - 4q_H)^2} \, , \tag{11.42}$$

and those referred to firm L:

$$\dot{a}_L = a_L (\delta + \rho) - \frac{\Theta q_H q_L}{4 (4q_H - q_L)} \; ; \tag{11.43}$$

$$\dot{b}_L = b_L (\delta + \rho) - \frac{\Theta^2 q_H (2q_H + q_L)}{4 (q_L - 4q_H)^2} \, . \tag{11.44}$$

The dynamic system formed by (11.41), (11.42), (11.43) and (11.44), together with the state equations (11.10) and (11.11), yields the following admissible steady state point:[11]

$$a_H^{CL} = 5.478\delta^2 (\delta + \rho) \; ; \; a_L^{CL} = 0.73208\delta^2 (\delta + \rho) \; ; \tag{11.45}$$

[11] As for the open-loop case, we find four equilibria, only the one reported in the text being admissible.

$$b_H^{CL} = 4.7735\delta^2(\delta+\rho)\ ;\ b_L^{CL} = 1.7166\delta^2(\delta+\rho)\ ; \tag{11.46}$$

$$\Theta^{CL} = 6.21\delta(\delta+\rho)\ ;\ q_H^{CL} = 4.7735\delta(\delta+\rho)\ ;\ q_L^{CL} = 1.7166\delta(\delta+\rho)\ . \tag{11.47}$$

Proposition 11.3 *The steady state identified by*

$$\{a_H^{CL}, a_H^{CL}, b_H^{CL}, b_L^{CL}, \Theta^{CL}, q_H^{CL}, q_L^{CL}\}$$

is saddle point equilibrium.

Proof. See Appendix 11.2. ■

By a direct comparison between equilibrium qualities and investments we can write:

Lemma 11.3 *Under the closed-loop memoryless solution concept, firm H invests more than firm L both in advertising and in quality improvement.*

Equilibrium output and price levels turn out to be:

$$x_H^{CL} = 2.7983\delta(\delta+\rho)\ ;\ x_L^{CL} = 1.7059\delta(\delta+\rho)\ ; \tag{11.48}$$

$$p_H^{CL} = 13.357\delta^2(\delta+\rho)^2\ ;\ p_L^{CL} = 2.9282\delta^2(\delta+\rho)^2\ , \tag{11.49}$$

provided that $x_H^{CL} + x_L^{CL} < 1$. Partial market coverage prevails iff:

$$\rho < \overline{\overline{\rho}} = -2.0\times10^{-10}\frac{-1.1101\times10^9 + 5.0\times10^9\delta^2}{\delta}\ . \tag{11.50}$$

Lemma 11.4 *Under the closed-loop memoryless solution concept, firm H attains a larger market share and charges a higher market price than firm L.*

As in the previous case, we may assess the relative performance of firms in terms of equilibrium profits:

$$\pi_H^{CL} = 37.377\delta^3(\delta+\rho)^3 - 52.795\delta^4(\delta+\rho)^2\ ; \tag{11.51}$$

$$\pi_L^{CL} = 4.9952\delta^3(\delta+\rho)^3 - 3.4827\delta^4(\delta+\rho)^2\ . \tag{11.52}$$

From a direct comparison between (11.51) and (11.52) we obtain:

Proposition 11.4 *Provided that $\rho \in \left[0, \overline{\overline{\rho}}\right)$, ensuring that partial market coverage prevails, under the closed-loop memoryless solution concept the following holds:*

(i) for all $\rho \in [0, 0.41256)$, *we have* $\pi_L^{CL} > 0 > \pi_H^{CL}$;

(ii) for all $\rho \in (0.41256, 0.52282\delta]$, *we have* $\pi_L^{CL} > \pi_H^{CL} > 0$;

(iii) for all $\rho > 0.52282\delta$, *we have* $\pi_H^{CL} > \pi_L^{CL} > 0$.

The above proposition is qualitatively equivalent to proposition 11.2. Therefore, also in the closed-loop case, there exists a range of parameters wherein the low quality firm performs better than the high quality firm in terms of equilibrium profits. Furthermore, in contrast with the so-called *finiteness property* (Shaked and Sutton, 1983), the firm providing the market with the inferior variety may become a natural monopolist.[12]

As to the comparison between open- and closed-loop equilibria, it is self-evident from (11.31) and (11.47) that the marginal willingness to pay and the level of product quality for both varieties are higher in the closed-loop case. Moreover,

Corollary 11.1 *The degree of vertical differentiation is larger at the closed-loop equilibrium than at the open-loop one.*

To see this, it suffices to carry out a straightforward computation, from which we get $q_L^{OL}/q_H^{OL} = 0.2929$ and $q_L^{CL}/q_H^{CL} = 0.35961$. So, the 'closed-loop motive', whereby firms explicitly take into consideration the rivals' behaviour as the game unravels over time, translates here into a wider product variety.

However, unlike the conventional wisdom according to which firms make greater investment efforts under closed-loop than under open-loop strategies (see Reynolds, 1987, *inter alia*), the firm providing the inferior quality makes a lower effort in the advertising activity when feedback effects are taken into account. This result depends on the fact that the benefits from any increase in consumers' marginal willingness to pay cannot be internalised, and spill over to the other firm. Indeed, the high quality firm invests much more in closed- than in open-loop yielding greater positive externalities to the rival. Equilibrium sales and market prices are always higher at the closed-loop equilibrium. A non-trivial result is that equilibrium profits are also higher in the closed-loop case than in the open-loop one. This is due to the fact that higher efforts (at least in quality supply) are adequately counterbalanced by larger revenues brought about by the increase in the marginal willingness to pay.

[12] In the differential game model at stake, contrary to what we have seen in the previous chapter, the so-called *4/7 rule* (Choi and Shin, 1992) never obtains, under either solution concept. Of course, this is due to the fact that firms behave *à la* Cournot. It is worth noting, however, that in Choi and Shin's model the quantity-setting behaviour would entail product homogeneity, with both firms supplying the highest feasible quality.

11.4 Concluding Remarks

We have investigated a differential duopoly game where each firm, through capital accumulation over time, may invest both in persuasive advertising campaigns aimed at increasing the willingness to pay of consumers and in an R&D process aimed at increasing the level of own product quality. The willingness to pay of consumers and the levels of product qualities have been treated as state variables evolving (in continuous time) in response to the interplay between firms' investments and decay rates. Unlike multi-stage games, differential games are particularly suitable to shed light on the nature of investments, which is inherently a dynamic one. To the best of our knowledge, the model presented in this chapter represents the first attempt to capture formally the dynamic incentives for oligopolistic firms to devote resources to advertising campaigns and quality improvements, jointly.

The main result we have obtained is that, in contrast with the acquired wisdom based on static models, the firm providing the market with the inferior variety may earn higher profits than the rival. Furthermore, we have shown that there exists a range of parameters wherein the low quality firm commands monopoly power. The rationale for these results is that, while in a static setting there are only instantaneous production costs, and we know from the existing static literature that it is always more profitable to produce the superior variety, in our dynamic setting things are much more involved: quality production and improvement require firms to make increasing efforts in each point in time, due to the assumption on dynamic decreasing returns w.r.t. the investment technology. Indeed, in the long run, it may become too costly for firms not only to produce but also to maintain a high quality level. This occurs when the discount rate is very small compared to the decay rate, i.e. future profits matter almost as present ones. When, instead, future profits are highly discounted, we come back to a *quasi*-static world where the relevant time horizon is perceived as a very short one due to the myopic attitude of firms, thus confirming the conventional static wisdom.

Appendices

Appendix 11.1: Proof of Proposition 11.1

Here we are interested in the stability properties of the open-loop dynamic system formed by (11.25), (11.26), (11.27) and (11.28) and the state equations (11.10) and (11.11). To verify that the steady state point defined in Proposition 11.1 is stable along a saddle path, we consider the following 7×7

matrix:

$$\Omega^{OL} = \begin{bmatrix} \frac{\partial \dot{\Theta}}{\partial \Theta} & \frac{\partial \dot{\Theta}}{\partial q_H} & \frac{\partial \dot{\Theta}}{\partial q_L} & \frac{\partial \dot{\Theta}}{\partial a_H} & \frac{\partial \dot{\Theta}}{\partial a_L} & \frac{\partial \dot{\Theta}}{\partial b_H} & \frac{\partial \dot{\Theta}}{\partial b_L} \\ \frac{\partial \dot{q}_H}{\partial \Theta} & \frac{\partial \dot{q}_H}{\partial q_H} & \frac{\partial \dot{q}_H}{\partial q_L} & \frac{\partial \dot{q}_H}{\partial a_H} & \frac{\partial \dot{q}_H}{\partial a_L} & \frac{\partial \dot{q}_H}{\partial b_H} & \frac{\partial \dot{q}_H}{\partial b_L} \\ \frac{\partial \dot{q}_L}{\partial \Theta} & \frac{\partial \dot{q}_L}{\partial q_H} & \frac{\partial \dot{q}_L}{\partial q_L} & \frac{\partial \dot{q}_L}{\partial a_H} & \frac{\partial \dot{q}_L}{\partial a_L} & \frac{\partial \dot{q}_L}{\partial b_H} & \frac{\partial \dot{q}_L}{\partial b_L} \\ \frac{\partial \dot{a}_H}{\partial \Theta} & \frac{\partial \dot{a}_H}{\partial q_H} & \frac{\partial \dot{a}_H}{\partial q_L} & \frac{\partial \dot{a}_H}{\partial a_H} & \frac{\partial \dot{a}_H}{\partial a_L} & \frac{\partial \dot{a}_H}{\partial b_H} & \frac{\partial \dot{a}_H}{\partial b_L} \\ \frac{\partial \dot{a}_L}{\partial \Theta} & \frac{\partial \dot{a}_L}{\partial q_H} & \frac{\partial \dot{a}_L}{\partial q_L} & \frac{\partial \dot{a}_L}{\partial a_H} & \frac{\partial \dot{a}_L}{\partial a_L} & \frac{\partial \dot{a}_L}{\partial b_H} & \frac{\partial \dot{a}_L}{\partial b_L} \\ \frac{\partial \dot{b}_H}{\partial \Theta} & \frac{\partial \dot{b}_H}{\partial q_H} & \frac{\partial \dot{b}_H}{\partial q_L} & \frac{\partial \dot{b}_H}{\partial a_H} & \frac{\partial \dot{b}_H}{\partial a_L} & \frac{\partial \dot{b}_H}{\partial b_H} & \frac{\partial \dot{b}_H}{\partial b_L} \\ \frac{\partial \dot{b}_L}{\partial \Theta} & \frac{\partial \dot{b}_L}{\partial q_H} & \frac{\partial \dot{b}_L}{\partial q_L} & \frac{\partial \dot{b}_L}{\partial a_H} & \frac{\partial \dot{b}_L}{\partial a_L} & \frac{\partial \dot{b}_L}{\partial b_H} & \frac{\partial \dot{b}_L}{\partial b_L} \end{bmatrix}_{q_i^{OL}, \Theta^{OL}}$$

By computing the seven eigenvalues, we find that at least one is negative over the entire admissible range of parameters:

$$\underline{\xi} = 0.5\rho - 0.3218\sqrt{8\delta^2 + 8\delta\rho + 2.4142\rho^2} < 0$$

while at least one is positive:

$$\overline{\xi} = \delta + \rho > 0$$

Therefore, the steady state $\{a_H^{OL}, a_H^{OL}, b_H^{OL}, b_L^{OL}, \Theta^{OL}, q_H^{OL}, q_L^{OL}\}$ is indeed a saddle point equilibrium. ■

Appendix 11.2: Proof of Proposition 11.3

In this case, we are interested in the closed-loop dynamic system formed by (11.41), (11.42), (11.43) and (11.44) and the state equations (11.10) and (11.11). To verify that the steady state point defined in Proposition 11.3 is stable along a saddle path, we consider the following 7×7 matrix (numerical values are approximate):

$$\Omega^{CL} = \begin{bmatrix} \frac{\partial\dot{\Theta}}{\partial\Theta} & \frac{\partial\dot{\Theta}}{\partial q_H} & \frac{\partial\dot{\Theta}}{\partial q_L} & \frac{\partial\dot{\Theta}}{\partial a_H} & \frac{\partial\dot{\Theta}}{\partial a_L} & \frac{\partial\dot{\Theta}}{\partial b_H} & \frac{\partial\dot{\Theta}}{\partial b_L} \\ \frac{\partial\dot{q}_H}{\partial\Theta} & \frac{\partial\dot{q}_H}{\partial q_H} & \frac{\partial\dot{q}_H}{\partial q_L} & \frac{\partial\dot{q}_H}{\partial a_H} & \frac{\partial\dot{q}_H}{\partial a_L} & \frac{\partial\dot{q}_H}{\partial b_H} & \frac{\partial\dot{q}_H}{\partial b_L} \\ \frac{\partial\dot{q}_L}{\partial\Theta} & \frac{\partial\dot{q}_L}{\partial q_H} & \frac{\partial\dot{q}_L}{\partial q_L} & \frac{\partial\dot{q}_L}{\partial a_H} & \frac{\partial\dot{q}_L}{\partial a_L} & \frac{\partial\dot{q}_L}{\partial b_H} & \frac{\partial\dot{q}_L}{\partial b_L} \\ \frac{\partial\dot{a}_H}{\partial\Theta} & \frac{\partial\dot{a}_H}{\partial q_H} & \frac{\partial\dot{a}_H}{\partial q_L} & \frac{\partial\dot{a}_H}{\partial a_H} & \frac{\partial\dot{a}_H}{\partial a_L} & \frac{\partial\dot{a}_H}{\partial b_H} & \frac{\partial\dot{a}_H}{\partial b_L} \\ \frac{\partial\dot{a}_L}{\partial\Theta} & \frac{\partial\dot{a}_L}{\partial q_H} & \frac{\partial\dot{a}_L}{\partial q_L} & \frac{\partial\dot{a}_L}{\partial a_H} & \frac{\partial\dot{a}_L}{\partial a_L} & \frac{\partial\dot{a}_L}{\partial b_H} & \frac{\partial\dot{a}_L}{\partial b_L} \\ \frac{\partial\dot{b}_H}{\partial\Theta} & \frac{\partial\dot{b}_H}{\partial q_H} & \frac{\partial\dot{b}_H}{\partial q_L} & \frac{\partial\dot{b}_H}{\partial a_H} & \frac{\partial\dot{b}_H}{\partial a_L} & \frac{\partial\dot{b}_H}{\partial b_H} & \frac{\partial\dot{b}_H}{\partial b_L} \\ \frac{\partial\dot{b}_L}{\partial\Theta} & \frac{\partial\dot{b}_L}{\partial q_H} & \frac{\partial\dot{b}_L}{\partial q_L} & \frac{\partial\dot{b}_L}{\partial a_H} & \frac{\partial\dot{b}_L}{\partial a_L} & \frac{\partial\dot{b}_L}{\partial b_H} & \frac{\partial\dot{b}_L}{\partial b_L} \end{bmatrix}_{q_i^{CL},\Theta^{CL}} =$$

$$\begin{bmatrix} -\delta & 0 & 0 & 1 & 1 & 0 & 0 \\ 0 & -\delta & 0 & 0 & 0 & 1 & 0 \\ 0 & 0 & -\delta & 0 & 0 & 0 & 1 \\ -0.88\delta(\delta+\rho) & -1.54\delta(\delta+\rho) & 1.08\delta(\delta+\rho) & \delta+\rho & 0 & 0 & 0 \\ -0.12\delta(\delta+\rho) & 0.02\delta(\delta+\rho) & -0.47\delta(\delta+\rho) & 0 & \delta+\rho & \delta+\rho & \delta+\rho \\ -1.54\delta(\delta+\rho) & -0.022\delta(\delta+\rho) & 0.06\delta(\delta+\rho) & 0 & 0 & \delta+\rho & 0 \\ -0.55\delta(\delta+\rho) & 0.13\delta(\delta+\rho) & -0.35\delta(\delta+\rho) & 0 & 0 & 0 & \delta+\rho \end{bmatrix}$$

Again, computing the seven eigenvalues we find that at least one is negative and at least one is positive over the entire admissible parameter range. Therefore, the steady state equilibrium is a saddle point. We omit the expressions of eigenvalues since they are cumbersome. Anyway, they are available upon request. ■

References

1] Case, J. (1979), *Economics and the Competitive Process*, New York, New York University Press.

2] Cellini, R. and L. Lambertini (2003), "Advertising in a Differential Oligopoly Game", *Journal of Optimization Theory and Applications*, **116**, 61-81.

3] Choi, J.C. and H.S. Shin (1992), "A Comment on a Model of Vertical Product Differentiation", *Journal of Industrial Economics*, **40**, 229-31.

4] Colombo, L. and L. Lambertini (2003), "Dynamic Advertising under Vertical Product Differentiation", *Journal of Optimization Theory and Applications*, 119, 261-80.

5] Conrad, K. (1985), "Quality, Advertising and the Formation of Goodwill under Dynamic Conditions", in G. Feichtinger (ed.), *Optimal Control Theory and Economic Analysis*, vol. 2, Amsterdam, North-Holland, 215-34.

6] Dockner, E.J., G. Feichtinger and S. Jørgensen (1985), "Tractable Classes of Nonzero-Sum Open-Loop Nash Differential Games: Theory and Examples", *Journal of Optimization Theory and Applications*, **45**, 179-97.

7] Dockner, E.J., S. Jørgensen, N.V. Long and G. Sorger (2000), *Differential Games in Economics and Management Science*, Cambridge, Cambridge University Press.

8] Dutta, P.K., S. Lach and A. Rustichini (1995), "Better Late than Early: Vertical Differentiation in the Adoption of a New Technology", *Journal of Economics and Management Strategy*, **4**, 563-89.

9] Erickson, G.M. (1991), *Dynamic Models of Advertising Competition*, Dordrecht, Kluwer.

10] Evans, G.C. (1924), "The Dynamics of Monopoly", *American Mathematical Monthly*, **31**, 75-83.

11] Feichtinger, G., R.F. Hartl and P.S. Sethi (1994), "Dynamic Optimal Control Models in Advertising: Recent Developments", *Management Science*, **40**, 195-226.

12] Fershtman, C. (1987), "Identification of Classes of Differential Games for which the Open-Loop is a Degenerate Feedback Nash Equilibrium", *Journal of Optimization Theory and Applications*, **55**, 217-31.

13] Gabszewicz, J.J. and J.-F. Thisse (1979), "Price Competition, Quality and Income Disparities", *Journal of Economic Theory*, **20**, 340-59.

14] Gabszewicz, J.J. and J.-F. Thisse (1980), "Entry (and Exit) in a Differentiated Industry", *Journal of Economic Theory*, **22**, 327-38.

15] Galbraith, J.K. (1967), *The New Industrial State*, New York, The New American Library.

16] Jørgensen, S. (1982), "A Survey of Some Differential Games in Advertising", *Journal of Economic Dynamics and Control*, **4**, 341-69.

17] Klemperer, P. (1995), "Competition when Consumers Have Switching Costs: An Overview with Applications to Industrial Organization, Macroeconomics, and International Trade", *Review of Economic Studies*, **62**, 515-39.

18] Kotowitz, Y. and F. Mathewson (1979), "Advertising, Consumer Information, and Product Quality", *Bell Journal of Economics*, **10**, 566-88.

19] Lehmann-Grube, U. (1997), "Strategic Choice of Quality when Quality is Costly: The Persistence of the High-Quality Advantage", *RAND Journal of Economics*, **28**, 372-84.
20] Leitmann, G. and W.E. Schmitendorf (1978), "Profit Maximization through Advertising: A Nonzero Sum Differential Game Approach", *IEEE Transactions on Automatic Control*, **23**, 646-50.
21] Mehlmann, A. and R. Willing (1983), "On Nonunique Closed-Loop Nash Equilibria for a Class of Differential Games with a Unique and Degenerate Feedback Solution", *Journal of Optimization Theory and Applications*, **41**, 463-72.
22] Moorthy, K.S. (1988), "Product and Price Competition in a Duopoly Model", *Marketing Science*, **7**, 141-68.
23] Motta, M. (1993), "Endogenous Quality Choice: Price vs Quantity Competition", *Journal of Industrial Economics*, **41**, 113-32.
24] Mussa, M., and S. Rosen (1978), "Monopoly and Product Quality," *Journal of Economic Theory*, **18**, 301-17.
25] Ouardighi, F.E. and F. Pasin (2002), "Advertising and Quality Decisions under Dynamic Conditions", X International Symposium on Dynamic Games and Applications, State University of St. Petersburg, July 8-11, 2002.
26] Reinganum, J. (1982), "A Class of Differential Games for Which the Closed Loop and Open Loop Nash Equilibria Coincide", *Journal of Optimization Theory and Applications*, **36**, 253-62.
27] Reynolds, S.S. (1987), "Capacity Investment, Preemption and Commitment in an Infinite Horizon Model", *International Economic Review*, **28**, 69-88.
28] Ringbeck, J. (1985), "Mixed Quality and Advertising Strategies under Asymmetric Information", in G. Feichtinger (ed.), *Optimal Control Theory and Economic Analysis*, vol. 2, Amsterdam, North-Holland, 197-214.
29] Schmalensee, R. (1978), "A Model of Advertising and Product Quality", *Journal of Political Economy*, **86**, 1213-25.
30] Sethi, S.P. (1977) "Dynamic Optimal Control Models in Advertising: A Survey", *SIAM Review*, **19**, 685-725.
30] Shaked, A. and J. Sutton (1982), "Relaxing Price Competition through Product Differentiation", *Review of Economic Studies*, **69**, 3-13.
31] Shaked, A. and J. Sutton (1983), "Natural Oligopolies", *Econometrica*, **51**, 1469-83.
32] Sorger, G. (1989), "Competitive Dynamic Advertising: A Modification of the Case Game", *Journal of Economic Dynamics and Control*, **13**, 55-80.
33] Vidale, M.L. and H.B. Wolfe (1957), "An Operations Research Study of Sales Response to Advertising", *Operations Research*, **5**, 370-81.

Index